RADIATION
Doses, Effects, Risks

 UNITED NATIONS ENVIRONMENT PROGRAMME

United Nations publication
Sales No. E.86.III.D.4
01000P

ISBN 92 807 1104 0

Design and artwork by Diagram Visual Information Limited, London.
Typesetting by Bournetype, Bournemouth.

Contents

Foreword

The early 1950s saw major concern developing in all countries about the effects of ionizing radiation. Not only was the horror of the Hiroshima and Nagasaki bombings fresh in everybody's memory, but three countries, by testing new nuclear devices in the atmosphere, had started to spread radioactive material world-wide. The effects of such radioactive fall-out were largely unknown and speculation was rife about the health consequences of that widespread exposure to radiation.

To meet that concern, in December 1955, the United Nations General Assembly established as one of its subsidiary bodies the Scientific Committee on the Effects of Atomic Radiation. The resolution establishing the Committee was a bold text both for what it said and for what it did not say. Rather than confining the task of the Committee to the study of fall-out, the issue that was then in everybody's mind, it did ask the Committee to review levels, effects and risks from all sources of radiation, both natural and man-made, including fall-out. It did not ask the committee to suggest remedies or make recommendations on action to be taken – merely to assess the current situation, untrammelled by responsibilities leading to decision-making.

Thirty years and eight massive reports later, the Committee still provides one of the few examples of how a soundly established body can accomplish consistently outstanding work of great value both to the scientific community, which keeps referring to these reports as the ultimate and most authoritative source of radiation data and assessments, and to the political community, which has found in the reports the solid factual basis leading to such instruments as the Partial Test Ban Treaty.

The booklet that I have the pleasure of introducing is published on the 30th anniversary of the Committee and aims at making the findings of the Committee available to a broader audience than have had access to them so far. In a field as complex and controversial as radiation, the use of technical jargon is unavoidable. I am grateful to the editor, and to a number of scientists who have assisted him, for keeping the level of technical language accessible to the general educated reader. Of course, it may not be easy reading, but the effort spent in mastering its complexities will repay the reader by enabling him to understand, and intelligently participate in, one of the great debates of our time.

Mostafa Kamal Tolba
Executive Director,
United Nations Environment Programme

Nairobi, December 1985

Introduction

Few scientific issues arouse so much public controversy as the effects of radiation. Scarcely a week seems to go by in developed countries without some expression of public feeling – and, as some developing countries advance their nuclear programmes, they may well increasingly have the same experience. There is little sign that the radiation debate will die down in the near future.

Unfortunately, providing unbiased factual information to the public often takes second place to propagating opinions. Too often, anti-nuclear activists rely on emotion; too often, nuclear advocates rely on bland reassurance.

The United Nations Scientific Committee on the Effects of Atomic Radiation (UNSCEAR) collects available evidence on the sources and effects of radiation, and evaluates it. It considers a wide range of natural and man-made sources, and its conclusions may surprise even some who have followed the public debate quite closely.

Radiation does kill. It causes severe tissue damage at high doses. At low levels it can cause cancers and induce genetic defects that can affect the children, grandchildren and later descendants of those irradiated.

But the most important sources of radiation to the general public are not those that attract the greatest attention. Natural sources contribute most exposure. Nuclear power contributes only a small proportion of the radiation emitted by human activities; much less controversial activities such as the use of x rays in medicine provide much greater doses. And other everyday activities like burning coal, air travel and – particularly – living in well-insulated homes can substantially increase exposure to natural radiation. The greatest causes for concern and the greatest scope for reducing human exposure to radiation lie in some of these uncontroversial pursuits, and are largely ignored by the debate.

This booklet does not pretend to have all the answers. Our knowledge is still inadequate, even though more is known about the sources, effects and risks of radiation than about those of almost any other toxic agent. But the booklet does try to summarise what solid information there is, so as to guide the debate onto firmer ground.

UNSCEAR was set up by the UN General Assembly in 1955 to evaluate doses, effects and risks from radiation on a world-wide scale. It brings together leading scientists from 20 countries and is one of the most authoritative bodies of its kind in the world. It does not set, or even recommend, safety standards; rather it provides information on radiation which enables such bodies as the International Commission on Radiological Protection and national authorities to do so. Every few years it produces major reports assessing, in considerable detail, the doses, effects and risks from all sources to which man is exposed. This booklet is an attempt to summarize the most up-to-date material from the reports for the general reader. But it is no substitute for the reports themselves.

The next four chapters are based on UNSCEAR's most recent reports to the United Nations' General Assembly, but they have not been reviewed or approved by the Committee. The last chapter is an attempt to discuss some general issues about the acceptability of risks from radiation, which are not part of the Committee's remit, and have never been covered in its reports.

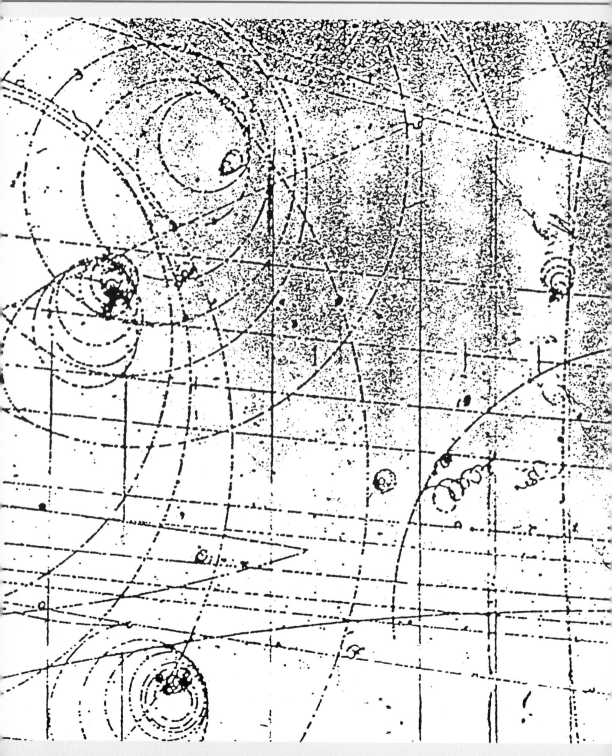

Radiation and Life

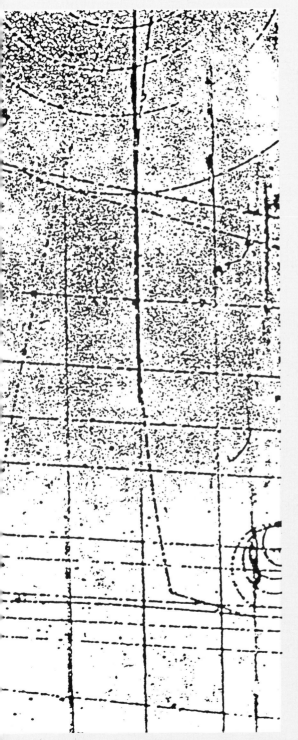

There is nothing new about radioactivity except the uses to which people have been learning to put it. Both radioactivity and the radiation it produces existed on earth long before life emerged. Indeed they were present in space before the earth itself appeared.

Radiation took part in the big bang which, as far as we know, gave birth to the universe about 20 billion years ago. Since then it has pervaded the cosmos. Radioactive materials became part of the earth at its very formation. Even man himself is slightly radioactive, for all living tissue contains traces of radioactive substances. But it is less than a century since humanity first discovered this elemental, universal phenomenon.

In 1896 Henri Becquerel, a French scientist, put some photographic plates away in a drawer, weighting them with bits of a mineral containing uranium. When he developed them he found, to his surprise, that they had been affected by radiation, and he traced this to the uranium. Soon afterwards, a young Polish-born chemist, Marie Curie, took the research further and was the first to coin the word "radioactivity". In 1898, she and her husband, Pierre, discovered that, as uranium gave off radiation, it mysteriously turned into other elements, one of which they called polonium, after her homeland, and another, radium, the "shining" element. Both Becquerel's and the Curies' work was greatly assisted by an earlier scientific breakthrough, when, in 1895 – and also by chance – Wilhelm Roentgen, a German physicist, discovered x rays.

It was not long before Becquerel experienced the most troublesome drawback of radiation, the effect it can have on living tissues. He put a vial of radium in his pocket and damaged his skin. Marie Curie was to die of a malignant blood disease probably – we now know – because of her exposure to radiation. At least 336 early radiation workers died from the doses they received.

Undeterred, a small group of brilliant, often young, scientists embarked on one of the most enthralling quests of all time, delving into the innermost secrets of matter itself. Their work was eventually to lead, in 1945, to the explosion of atomic bombs at the end of the Second World War with disastrous loss of life. It also led, in 1956, to the world's first sizeable nuclear power station, Calder Hall, in the United Kingdom. Meanwhile, ever since Roentgen's discoveries, there has been a continuous expansion of the medical uses of radiation.

The focus of the scientists' quest was the atom and, more particularly, its structure. We now know that atoms behave like miniature solar systems; tiny nuclei are surrounded by orbiting "planets" called electrons. The nucleus is only about one hundred thousandth of the size of the entire atom, but it is so dense that it accounts for almost all its mass. It is generally a cluster of particles which cling tightly to each other (see diagram 2.1).

Some of the particles carry a positive electrical charge and are called protons. The number of protons decide what element an atom belongs to; an atom of hydrogen has a single proton, an atom of oxygen eight, an atom of uranium 92. Each atom has the same number of orbiting electrons as it has protons; the electrons are negatively charged, and so they and the positively-charged protons balance each other. As a result, the atom itself is neither positive nor negative, but neutral.

The rest of the particles in the nuclear cluster are called neutrons because they carry no electrical charge. Atoms of the same element always have the same number of protons in their nuclei, but they can have differing numbers of neutrons. Those that have differing neutron numbers, but the same number of protons, belong to different varieties of the same element, and are called its "isotopes". These are distinguished by adding up the total numbers of particles in their nuclei. Thus uranium-238 has 92 protons and 146 neutrons; uranium-235 has the same 92 protons, but 143 neutrons. The atoms thus characterised are called "nuclides".

Some nuclides are stable, and lead uneventful, unchanging lives. But they are a minority. Most are unstable, and give vent to

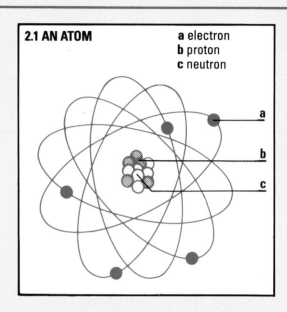

2.1 AN ATOM
a electron
b proton
c neutron

their instability by relentlessly trying to become something else. To take just one example, the particles in the nucleus of a uranium-238 atom are only just able to cluster together. Eventually a chunk of two protons and two neutrons will break away. As it goes, the uranium-238 turns into thorium-234 (with 90 protons and 144 neutrons). But thorium-234, too, is unstable; it, too, wants to become something else. It transforms itself by a different process; one of its neutrons turns into a proton, and it becomes protactinium-234, with 91 protons and 143 neutrons. When the proton metamorphoses, one of the orbiting electrons loses its partner and breaks away. Protactinium is extremely unstable and loses no time in changing its own shape, and so, by one of these means after the other the atom goes on transforming itself and shedding particles until it finally ends up as stable lead (diagram 2.3, overleaf). Of course, there are many such sequences of transformation, or "decay" as it is called, with a large variety of patterns and combinations.

As each change takes place, energy is released, and is transmitted as radiation. Loosely speaking, the emission of a chunk of two protons and two neutrons, as from

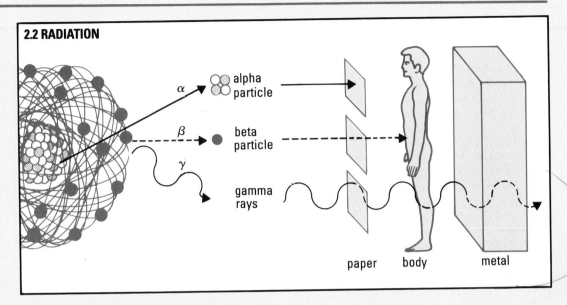

2.2 RADIATION

α — alpha particle

β — beta particle

γ — gamma rays

paper   body   metal

Three types of radiation and their penetrating power.

uranium-238, is "alpha" radiation; the emission of an electron, as from thorium-234, is "beta" radiation. Frequently the unstable nuclide will be so excited that the emission of particles is not sufficient to calm it down; it then gives off a vigorous burst of pure energy called "gamma" radiation. Like x rays (which are similar in many ways), gamma radiation does not involve the emission of particles.

The whole transformation process is called "radioactivity", and the unstable nuclides "radionuclides". But while all radionuclides are unstable, some are more unstable than others. Protactinium-234, for example, cannot wait to transform itself, while uranium-238 is extremely leisurely about the process. Half of a lump of protactinium-234 metamorphoses in little over a minute, whereas half a lump of uranium-238 will take four and a half billion years to turn into thorium-234. The period it takes half of any amount of an element to decay is known as its "half-life". The process goes on relentlessly. After one half-life, 50 out of 100 atoms will have remained unchanged; during the next half-life, 25 of these will decay, and so on exponentially. The number of transformations that take place each second

in an amount of radioactive material is called its "activity". The activity is measured in units called becquerels, after the man who discovered the phemonenon; each becquerel equals one transformation per second.

The different forms of radiation are emitted with different energies and penetrating power – and so have different effects on living things (diagram 2.2). Alpha radiation, with its heavy chunk of neutrons and protons is halted, for example, by a sheet of paper, and can scarcely penetrate the dead outer layers of the skin. So it is not dangerous unless substances emitting it get into the body through an open wound or are eaten or breathed in – but then it is especially damaging. Beta radiation is more penetrating. It will go through a centimetre or two of living tissue. Gamma radiation, which travels at the speed of light, is extremely penetrating and will go through anything short of a thick slab of lead or concrete.

It is the energy of radiation which does the damage, and the amount of energy deposited in living tissue is called the "dose" – a rather misleading term originally intended to remind people of doses of medicines. The dose may come from any radionuclide, or number of

2.3 RADIOACTIVE DECAY

type of radiation	nuclide	half-life
α	uranium–238	4.47 billion yrs.
β	thorium–234	24.1 days
β	protactinium–234	1.17 minutes
α	uranium–234	245000 years
α	thorium–230	8000 years
α	radium–226	1600 years
α	radon–222	3.823 days
α	polonium–218	3.05 minutes
β	lead–214	26.8 minutes
β	bismuth–214	19.7 minutes
α	polonium–214	0.000164 seconds
β	lead–210	22.3 years
β	bismuth–210	5.01 days
α	polonium–210	138.4 days
	lead–206	stable

The decay of uranium-238.

radionuclides, whether they remain outside the body or irradiate it from inside after being inhaled in air or swallowed in food or water. Doses are expressed in different ways depending on how much of the body, and what parts of it, are irradiated, whether one individual or many people are exposed, and the period during which the exposure takes place.

The amount of radiation energy that is absorbed per gram of tissue is called the absorbed **dose** (diagram 2.4) and is measured in units called grays (Gy). But this does not tell the full story because the same dose of alpha radiation is much more damaging than one of beta or gamma radiation. So the dose needs to be weighted for its potential to do damage, with alpha radiation given twenty times the weight of the others. This weighted dose is known as the "dose **equivalent**", and it is measured in units called sieverts (Sv) (diagram 2.5).

There is another refinement to be made. Some parts of the body are more vulnerable than others; a given dose equivalent of radiation is more likely to cause fatal cancer in the lung than in the thyroid, for example – and the reproductive organs are of particular concern because of the risk of genetic damage. The different parts of the body are therefore also given weightings (diagram 2.6). Once it has been weighted appropriately, the dose equivalent becomes the "**effective** dose equivalent", also expressed in sieverts.

This, however, describes only individual doses. If you add up all the individual effective dose equivalents received by a group of people, the result is called "the **collective** effective dose equivalent", and this is expressed in man-sieverts (man-Sv). But one further definition must be introduced, because many radionuclides decay so slowly that they are radioactive far into the future. This is the collective effective dose equivalent that will be delivered to generations of people over time, and it is called the "collective effective dose equivalent **commitment**".

This hierarchy of concepts may appear complicated, but it does bring them into a

2.4 DOSES

$\alpha \beta \gamma$

Absorbed Dose: *Energy imparted by radiation per gram of tissue.*

Dose equivalent: *Absorbed dose weighted for the potential of different radiations to do damage.*

Effective dose equivalent: *Dose equivalent weighted for the susceptibility to harm of different tissues.*

Collective effective dose equivalent: *Effective dose equivalent to a group of people from a source of radiation.*

Collective effective dose equivalent commitment: *Collective effective dose equivalent delivered over time to generations of people.*

coherent structure, and allows doses to be recorded consistently and comparably. To make things as simple as possible, the following chapters will avoid the use of these terms wherever possible. But frequently there is no alternative to them, to ensure accuracy and to eliminate ambiguity.

Risk weighting factors recommended by the International Commission on Radiological Protection for the calculation of effective dose equivalent.

2.5 UNITS

Becquerel (Bq): *The special name for the unit of **activity.** One becquerel corresponds to one disintegration per second of any radionuclide.*

Gray (Gy): *The special name for the unit of **absorbed dose.** It is the quantity of energy imparted by ionising radiation to a unit mass of matter such as tissue. One gray corresponds to one joule per kilogram.*

Sievert (Sv): *The special name of a unit of **dose equivalent.** This is the absorbed dose weighted according to the potential to do damage of the radiation that gives rise to it. One sievert also corresponds to one joule per kilogram.*

2.6 RISK WEIGHTING FACTORS

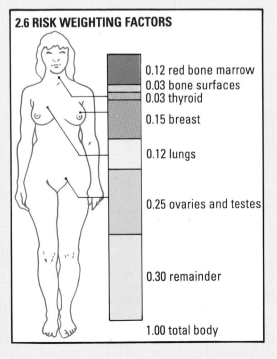

0.12 red bone marrow
0.03 bone surfaces
0.03 thyroid

0.15 breast

0.12 lungs

0.25 ovaries and testes

0.30 remainder

1.00 total body

3

Natural Sources 3

By far the greatest part of the radiation received by the world's population comes from natural sources (diagram 3.1, overleaf). Exposure to most of them is inescapable. Throughout the earth's history, radiation has fallen on its surface from outer space and risen from radioactive materials in its crust. People are irradiated in two ways. Radioactive substances may remain outside the body and irradiate it from the outside, or "externally". Or they may be inhaled in air or swallowed in food and water, and so irradiate people from inside, or "internally".

But though everyone on the planet receives natural radiation, some people get much more than others. This may result from where they live. Doses at some places, with particularly radioactive rocks or soils, are much higher than the average; at other places they are much less. Or it may result from their lifestyle. The use of particular building materials for houses, cooking with gas, open coal fires, home insulation, and even air travel all increase exposure to natural radiation.

Overall, terrestrial sources are responsible for most of man's exposure to natural radiation. In normal circumstances, they provide more than five-sixths of the annual effective dose equivalents received by individual people – most of it by internal irradiation. Cosmic rays contribute the remainder, mainly by external irradiation (diagram 3.2, overleaf).

This chapter looks first at external radiation from cosmic and terrestrial sources. It then considers internal radiation, paying particular attention to radon, a radioactive gas which is the biggest single source of average doses from natural radiation. Finally, it turns to several practices, from coal burning to the use of fertilizers, which release radioactive substances from the ground and so enhance man's exposure to terrestrial sources.

Cosmic rays
Just under half of man's exposure to external natural radiation comes from cosmic rays (diagram 3.2). Most of these originate from

3.1 SOURCES OF RADIATION

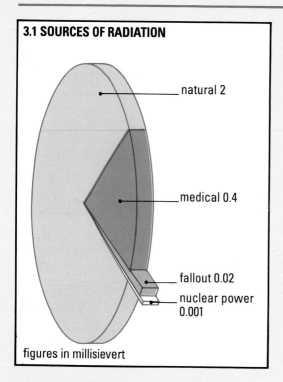

natural 2

medical 0.4

fallout 0.02

nuclear power 0.001

figures in millisievert

Average annual effective dose equivalents from **natural and man-made** *sources of radiation.*

3.2 NATURAL SOURCES

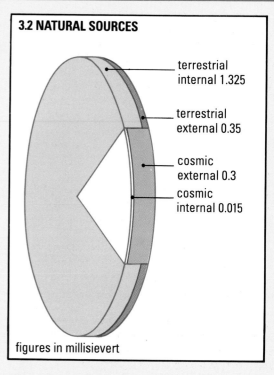

terrestrial internal 1.325

terrestrial external 0.35

cosmic external 0.3

cosmic internal 0.015

figures in millisievert

Average annual effective dose equivalents from **natural** *sources of radiation.*

deep in interstellar space; some are released from the sun during solar flares. They irradiate the earth directly, and interact with the atmosphere to produce further types of radiation and different radioactive materials.

Nowhere escapes this universal, invisible shower. But it affects some parts of the globe more than others. The poles receive more than the equatorial regions, because the earth's magnetic field diverts the radiation. But, more important, the level increases with altitude, since there is less air overhead to act as a shield.

3.3 AIR TRAVEL

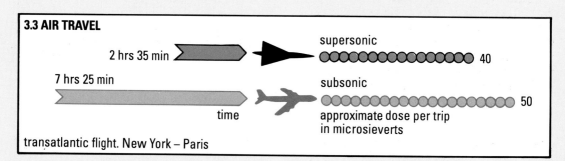

2 hrs 35 min

supersonic

40

7 hrs 25 min

subsonic

50

time

approximate dose per trip in microsieverts

transatlantic flight. New York – Paris

Calculated doses from cosmic rays to somebody flying the Atlantic in subsonic and supersonic aircraft (average solar conditions).

Someone living at sea level will, on average, receive an effective dose equivalent of about 300 microsievert (millions of a sievert) of cosmic radiation every year, but someone living above 2000 metres will receive several times as much. Air travel exposes passengers and crews to even higher dose rates, albeit for short periods at a time. Between 4000 metres, the altitude of the loftiest permanent Sherpa villages on the flanks of Mt. Everest, and 12,000 metres, the level of the highest intercontinental flights, exposures to cosmic radiation increase about twenty-five times. They rise further between 12,000 and 20,000 metres, the maximum altitude of supersonic aircraft (diagram 3.4).

A trip from New York to Paris will expose a passenger to about 50 microsievert in a normal jet aeroplane, and about 20 per cent less in a supersonic aircraft – although a supersonic aircraft is exposed to more intense radiation, it completes the journey much more quickly (diagram 3.3). In all, air travel results in a collective effective dose equivalent to the world's population of about 2000 man-sievert a year.

Terrestrial radiation

The main radioactive materials in rocks are potassium-40, rubidium-87, and two series of radioactive elements arising from the decay of uranium-238 and thorium-232, two long-lived radionuclides that have remained on earth since its origin.

Naturally, the levels of terrestrial radiation differ from place to place around the world, as the concentrations of these materials in the earth's crust vary. For most people, the variation is not particularly dramatic. Studies in France, the Federal Republic of Germany, Italy, Japan and the United States, for example, suggest that about 95 per cent of the people live in areas where the average dose rate varies from 0.3 to 0.6 millisievert (thousandths of a sievert) a year. But some receive much greater doses; about three per cent are exposed to one millisievert a year – half of them to over 1.4 millisievert a year. And there are places on earth where

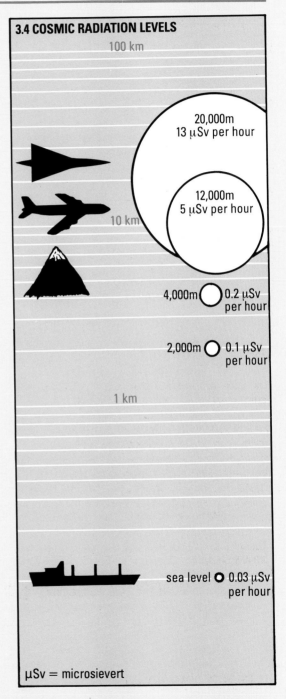

3.4 COSMIC RADIATION LEVELS

100 km

20,000m
13 µSv per hour

12,000m
5 µSv per hour

10 km

4,000m 0.2 µSv per hour

2,000m 0.1 µSv per hour

1 km

sea level 0.03 µSv per hour

µSv = microsievert

Increase with altitude of the dose equivalent rate from cosmic radiation. (Note that the diagram uses a logarithmic scale).

terrestrial radiation levels are very much higher still (diagram 3.5).

Near the city of Poços de Caldas, 200 kilometres north of Saō Paolo, Brazil, stands a small hill. Here researchers have discovered radiation dose rates up to about eight hundred times the average – 250 millisievert a year. As it happens, the hill is uninhabited. But only slightly less spectacular levels have been found in a coastal resort some 600 kilometres to the East.

Guarapari is a small town of 12,000 people, which every summer plays host to about 30,000 holiday makers. Particular spots on its beach have registered 175 millisievert a year. Radiation levels in the streets were found to be a good deal less, ranging from eight to 15 millisievert a year, but still many times higher than normal levels. It is much the same story in the fishing village of Meaipe, 50 kilometres to the south. Both stand on sands rich in thorium.

Half a world away on the south-west coast of India, 70,000 people live in a 55 kilometre long strip of land which also contains thorium-rich sands. Studies of 8,513 people in the area showed that they absorbed, on average, 3.8 millisievert of radiation a year. Over 500 of them received more than 8.7 millisievert a year. About sixty absorbed more than 17 millisievert – about 50 times the average dose from external terrestrial radiation.

These areas in Brazil and India are the best studied hot-spots on earth. But levels of up to 400 millisievert a year have been found in Ramsar, Iran, where there are springs rich in radium. And other regions of high terrestrial natural radiation are known to exist in France, Madagascar and Nigeria.

On average, UNSCEAR calculates that the world's people receive an effective dose equivalent of about 350 microsievert a year from external natural terrestrial radiation, slightly more than the average person living at sea level will receive from cosmic rays.

Internal irradiation

On average, two-thirds of the effective dose equivalent that people receive from natural sources comes from radioactive substances in the air they breathe, the food they eat and the water they drink.

Very little indeed of this internal dose comes from radioactive substances – like carbon-14 and tritium – which are formed by cosmic radiation. Almost all of it is derived from terrestrial sources. On average, people receive about 180 microsievert a year from potassium-40, which is absorbed in the body along with non-radioactive potassium, an essential element. But by far the greatest amount comes from the elements resulting from the decay of uranium-238 – and, to a lesser extent, from the decay of thorium-232.

Some of these, like lead-210 and polonium-210, mainly enter the body in food. Both become concentrated in fish and shellfish; people eating large amounts of seafood can expect to receive correspondingly high doses.

Tens of thousands of people in the extreme north of the northern hemisphere subsist mainly on reindeer (or caribou) meat. The animals contain high concentrations of these two radioactive materials, particularly polonium-210, because during the winter they graze on lichens which happen to accumulate them. The people end up with doses of polonium-210 up to 35 times normal levels. Meanwhile, at the other end of the world, people living in an uranium-rich area of Western Australia receive up to 75 times the normal dose from the sheep and kangaroo meat and offal they consume.

Radioactive substances like these often take complex routes through the environment before reaching man. Such routes, or "pathways", are often used to calculate doses received from particular sources. A simplified example of one set of pathways is shown in diagram 3.6 overleaf.

Radon

Scientists have only recently begun to realise that the most important of all sources of natural radiation is a tasteless, odourless, invisible gas, seven and a half times heavier than air, called radon. UNSCEAR now estimates that, together with its "daughters" – radionuclides formed as it decays – radon

3.5 TERRESTRIAL RADIATION, AND RADON
Some areas with high levels of terrestrial radiation.

A Poços de Caldas & Guarapari
B Kerala and Tamil Nadu
C Ramsar

Some measurements of activity concentrations of radon-222 in outdoor air in different parts of the world.

Bq per cubic metre

1 Cincinnati	9.6
2 France	9.3
3 New York City	4.8
4 United Kingdom	3.3
5 Washington	2.9
6 Japan	2.1
7 Bolivia	1.5
8 Philippines	0.3
9 Indian Ocean	0.07
10 Marianas	0.05
11 Marshall Islands	0.02
12 Caroline Islands	0.02

activity concentrations of radon–222 per cubic metre of air

average level about 2

● = 0.25 Bq per cubic metre

17

3.6 ENVIRONMENTAL PATHWAYS

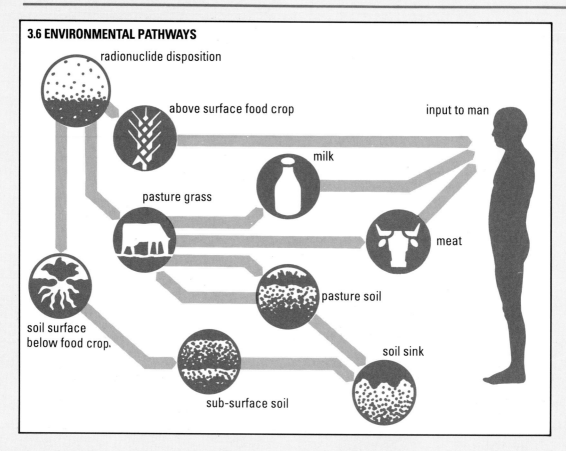

radionuclide disposition

above surface food crop

input to man

milk

pasture grass

meat

soil surface below food crop

pasture soil

sub-surface soil

soil sink

A set of environmental pathways. The diagram, based on a computer model used to calculate doses from the deposition of radionuclides from nuclear power production, shows both how they can reach man through the food he eats and how they can end up in the deeper layers of the soil. It is, necessarily, a simplification: highly complex processes are involved at almost every stage.

normally contributes about three-quarters of the annual effective dose equivalents received by individual people from terrestrial sources – and about half their doses from all natural sources put together. Most of the dose results from breathing in the radionuclides, particularly indoors.

In fact, radon has two main forms – radon-222, one of the radionuclides in the sequence formed by the decay of uranium-238, and radon-220, produced during the decay series of thorium-232. Radon-222 seems to be 20 times more important than radon-220, but for convenience both forms of the radioactive gas will be referred to here as "radon". Most of

the doses are in fact contributed by the radon daughters rather than by the gas itself.

Radon seeps out of the earth all over the world, but levels in outside air vary markedly from place to place (diagram 3,5). Perhaps paradoxically, however, people are mainly exposed to radon indoors. In temperate parts of the world the concentrations of radon indoors are, on average, about eight times higher than they are outside. There are no measurements in tropical countries, but because the weather is hotter, and so buildings are much more open, there is probably not much difference between indoor and outdoor concentrations.

Radon concentrates in indoor air when

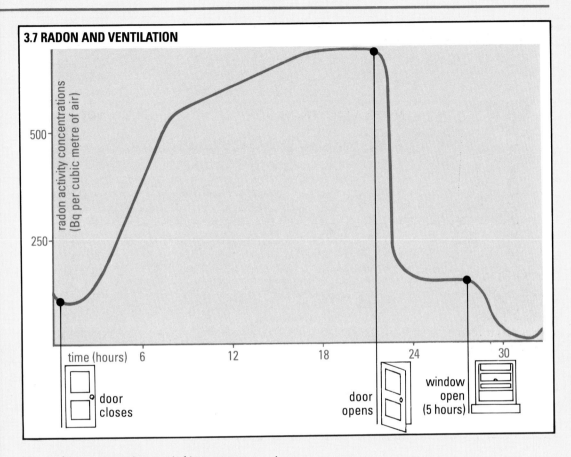

3.7 RADON AND VENTILATION

radon activity concentrations (Bq per cubic metre of air)

500

250

time (hours)   6   12   18   24   30

door
closes

door
opens

window
open
(5 hours)

How radon concentrations varied in one apartment house.

buildings are, by and large, closed spaces (diagram 3.7). Once the gas gets in, by filtering up through the floorboards from the ground, or, to a lesser extent, by seeping out of the very materials used to build the edifice, it finds it hard to get out. Very high levels of radiation can result, especially if the house, or other building, happens to stand on particularly radioactive ground, or has been erected with especially radioactive materials. And heavy insulation makes things much worse, by making it harder for the gas to escape.

Very high levels of radon are being found more and more frequently. In the late 1970s, concentrations more than 5000 times typical

levels in outside air were found in homes in Sweden and Finland. By the time of the latest UNSCEAR report, in 1982, homes with concentrations 500 times typical outdoor levels had been found in Great Britain and the United States – and, since then, dwellings have been discovered in both countries with radon levels equalling the highest found in Scandinavia. As the number of homes examined has increased, so has the number found to contain these extreme radon concentrations.

The commonest building materials – wood, bricks, and concrete – give off relatively little radon (diagram 3.8, overleaf). Granite is much more radioactive and so is pumice stone, used, for example, in the Soviet Union and West Germany. And some materials have given builders, scientists – and residents – unwelcome surprises by proving to be especially radioactive.

For several decades, for example, alum shales were used in the making of concrete in Sweden. The concrete was incorporated in between 350,000 and 700,000 Swedish homes. Then it was found that the shales were highly radioactive. Their use was cut back in the mid 1970s and then stopped completely. Calcium silicate slag – a highly radioactive by-product of the processing of phosphate ore – is used to make concrete and other building materials in North America. It has turned up in buildings in Idaho, Florida and Canada.

Phosphogypsum, another by-product from a different way of processing phosphate ore, has been widely used to make building blocks, plasterboard, partition systems and cement. It costs less than natural gypsum and was welcomed by environmentalists because it is a waste product, and so its use preserves natural resources and reduces pollution. In Japan alone, three million tonnes of the material were used in the construction industry in 1974. But it is also many times more radioactive than the natural gypsum it replaces, and people who live in houses containing it can expect to be exposed to about 30 per cent more radiation than those who do not. In all, it is estimated to give rise to a collective effective dose equivalent commitment of about 300,000 man-sieverts.

Other highly radioactive waste products used in building include red mud bricks from aluminium production; blast furnace slag from iron works; and fly ash from the burning of coal.

Even wastes from uranium mining have been used. Between 1952 and 1966 tailings from uranium mills were used as building materials and under houses, particularly in Grand Junction, Colorado. In Port Hope, Ontario, material from a radium recovery plant was used for construction. In both cases, the national governments had to step in and take remedial action because of the radiation doses received by inhabitants.

Despite all the concern about building materials, the ground underneath houses is almost always a greater source of radon. In some cases, houses have been built on old radioactive wastes, including uranium mine tailings in Colorado, alum shale tailings in Sweden, radium factory tailings in Australia, and reclaimed land from phosphate mining in Florida. But, even in more normal circumstances, most radiation comes up through the floor.

The highest radon levels in Helsinki, Finland, more than 5000 times greater than typical levels in outside air, were found in

Mean activity concentrations in building materials as measured in various countries.

3.8 BUILDING MATERIALS

wood (Finland)	1.1
natural gypsum (UK)	29
sand and gravel (FDR)	< 34
portland cement (FDR)	< 45
bricks (FDR)	126
granite (UK)	170
fly ash (FDR)	
alum shale (Sweden) 1974–1979	
alum shale (Sweden) 1929–1975	
phosphogypsum (FDR)	
calcium silicate slag (USA)	
uranium mine tailings (USA)	

homes where the only significant source can have been the ground on which they stood. Even in Sweden, with its difficulties arising from the use of alum shale, new research indicates that the greatest problem is radon emanating from the ground.

Radon concentrations in the upper storeys of high buildings tend to be lower than at the ground floor. A survey in Norway even showed that wooden houses had higher radon concentrations than brick ones, despite the fact that the wood gave off virtually none of the gas. This was because the wood houses generally had fewer storeys and so their rooms were closer to the radon-emitting ground.

The thickness and integrity of the floorings of buildings determines how much of the radon rising up from the ground actually gets in. Studies of houses built on reclaimed phosphate land in Florida have demonstrated this; while, in Chicago, houses built on bare earth, with unpaved crawl spaces, were found to have radon concentrations well over 100 times typical outdoor levels, even though the concentrations in the soil were normal.

By the same token, radon levels in houses can be reduced by sealing floors and walls. Experiments are still going on, but some promising results have already been achieved. Using fans to ventilate crawl spaces is a particularly effective way of reducing

the radon that gets in through the floor. Meanwhile covering walls with plastic materials like polyamide, polyvinylchloride, polyethylene and epoxy paint – or giving them three coats of oil-based paints – reduces the emission of radon from this source tenfold. Even wallpaper may cut it down by 30 per cent or so.

Water and natural gas provide further, if less important, sources of radon in homes (diagram 3.9, overleaf). Usually the amounts of radon in water are extremely small, but some supplies, especially from deep wells, have very high concentrations (diagram 3.10, overleaf). Such high levels have been found, for example, in wells in Finland and the United States, supplying water to Helsinki and, perhaps appropriately, to Hot Springs, Arkansas, among other places. The most radioactive water supplies have radon activity concentrations of one hundred million becquerel per cubic metre – the least, virtually nothing. In all, UNSCEAR estimates that less than one per cent of the world's population consumes water containing more than a million becquerel of radon activity per cubic metre, and less than ten per cent drink water with over 100,000 becquerel per cubic metre.

Oddly, perhaps, consuming water containing radon is not the main problem, even where levels are high. Generally, people

Bq of radium and
thorium per kilogram

341

496

1367

< 574

2140

4625

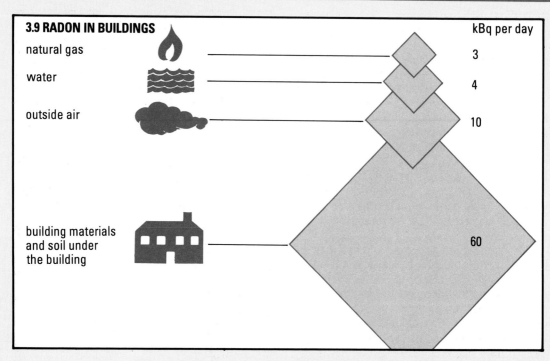

3.9 RADON IN BUILDINGS

		kBq per day
natural gas		3
water		4
outside air		10
building materials and soil under the building		60

The relative contribution of different radon sources in a reference house.

Average radon activity concentrations (kBq per cubic metre) in water sources.

take in most of their water in food and in hot drinks like tea and coffee. Boiling water, and cooking with it, releases most of the radon, and so the main intake comes from drinking the water cold. And even this is eliminated from the body very quickly.

People are more at risk from breathing in

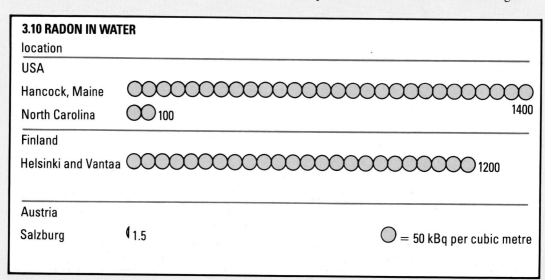

3.10 RADON IN WATER

location	
USA	
Hancock, Maine	1400
North Carolina	100
Finland	
Helsinki and Vantaa	1200
Austria	
Salzburg	1.5

○ = 50 kBq per cubic metre

the radioactivity emitted by radon-rich water – particularly in the bathroom. A survey in Finnish homes showed that, on average, radon concentrations in bathrooms were about three times higher than in kitchens, where less water was used, and some 40 times higher than in living rooms (diagram 3.12). Meanwhile, a study in Canada showed that the amount of radon and its daughters in bathroom air increased rapidly during a seven minute warm shower, and that it was well over an hour and a half after the shower was turned off before levels returned to anything like what they were originally (diagram 3.11).

Radon also gets into "natural gas" in the ground. Processing and storage removes much of it before the gas reaches the consumer, but radon concentrations in homes can still increase significantly if the gas is burned in unvented stoves, heaters and other appliances. If the appliances are vented outside the house, the increase is negligible.

Much of the radon removed from natural gas during processing ends up in liquefied petroleum gas (LPG), which is produced as a by-product. But natural gas provides ten to a hundred times more radiation to homes on a national basis than the more radioactive LPG because much more of it is burned.

Average radon activity concentrations in air caused by radon in water in a study of 20 Finnish houses.

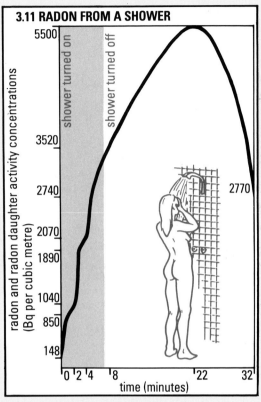

3.11 RADON FROM A SHOWER

radon and radon daughter activity concentrations (Bq per cubic metre)

shower turned on — shower turned off

5500
3520
2740
2770
2070
1890
1040
850
148

time (minutes)
0 2 4 8 22 32

Activity concentrations of radon and radon daughters in indoor air during and after a seven minute warm shower in a Canadian house. (The radon concentration in the water was 4,400 becquerel per cubic metre.)

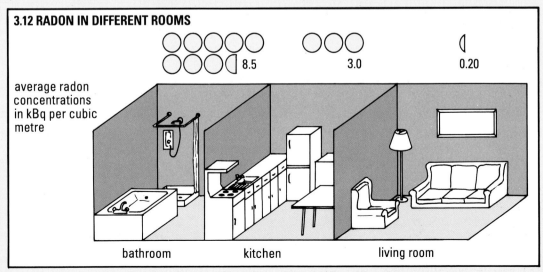

3.12 RADON IN DIFFERENT ROOMS

8.5   3.0   0.20

average radon concentrations in kBq per cubic metre

bathroom   kitchen   living room

Energy conservation measures can greatly increase radon concentrations. Insulating houses and stopping draughts reduces ventilation. This conserves heat, but also allows radon to build up.

Sweden, where houses are particularly heavily insulated, is especially affected. For many years radon in homes was not thought to be a problem in that country, despite the use of alum shales; a survey in 1956 suggested that there were no grounds for serious concern at the ventilation rates of the time. But since the early 1950s, ventilation rates in Swedish houses have been steadily dropping in a drive to save energy. Between 1950 and the mid 1970s ventilation rates were cut by more than half – and radon concentrations more than trebled (diagram 3.13). It has been calculated that every gigawatt year of electricity saved by reduced ventilation will expose Swedes to 5,600 man-sievert of extra radiation.

Tight energy conservation, high radon emissions from the ground to low rise buildings, and the use of alum shales in construction explain the Swedish situation. In 1982, UNSCEAR reported much more limited information from other countries which suggested that 90 per cent of their houses had radon daughter activity concentrations under 50 becquerel per cubic metre, about 25 times typical outdoor levels, and that no more than a few per cent contained amounts over 100 becquerel per cubic metre. In Sweden, by contrast, the same report noted, more than 30 per cent of the buildings were above this higher level, and average concentrations were four times those in other temperate lands.

There are, however, recent indications that Sweden may not be as exceptional as was once thought. Other countries are beginning to find that their problems are greater than they suspected. It is likely that part of the reason why Sweden appears to have a worse problem than elsewhere is that it carried out more extensive surveys earlier than in other countries.

The proportion of houses with radon daughter activity concentrations levels between 1000 and 10,000 becquerel per cubic metre ranges from 0.01% to 0.1% in different countries. This means that quite large numbers of people may be exposed to high concentrations in their homes. Nevertheless, in countries with problems less acute than in Sweden, three-quarters of the total collective dose equivalent will be accounted for by the homes with concentrations below 100 becquerel per cubic metre. The total effective dose equivalent due to exposure to radon and its daughters is normally about one millisievert a year – or about a half of the total estimated dose from all natural sources of radiation.

Other sources

Coal, like most natural materials, contains traces of primordial radionuclides. Burning it releases these, once locked deep in the earth, into the environment where they can affect people.

Though concentrations can vary a hundredfold from seam to seam, most coal contains lesser amounts of radioactive materials than the average in the earth's crust. But when coal is burned, most of its mineral matter is fused into ash, and most of the radioactive substances are concentrated in the ash as well. Most of the ash is heavy and drops to the bottom of a power plant furnace. Lighter fly ash, however, is carried up the chimney of the plant. How much comes out depends on how much of an attempt is made to stop it with anti-pollution devices.

The cloud from the chimneys irradiates people and the pollution also settles on the ground and contaminates food crops. Some of it may even get back into the air as dust. The production of each gigawatt year of electrical energy is estimated to lead to a total collective dose equivalent commitment of two man-sievert, which means that in 1979 the world's coal power stations produced a collective effective dose equivalent commitment of about 2000 man-sievert.

Less coal is used for cooking and heating in private homes, but a greater proportion of the ash escapes. So the world's stoves

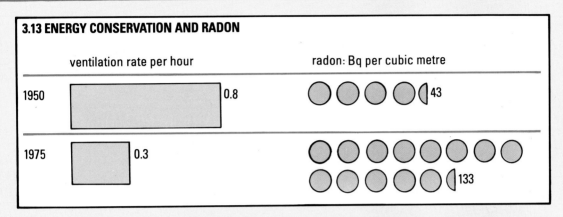

3.13 ENERGY CONSERVATION AND RADON

	ventilation rate per hour	radon: Bq per cubic metre
1950	0.8	43
1975	0.3	133

Decrease of ventilation rates and increase of average radon activity concentrations in Swedish apartment houses.

and fireplaces may well emit as much ash to the atmosphere as its power stations. Furthermore, unlike most power stations, private homes have low chimneys and are normally in the heart of population centres; so much more of the pollution will fall on people. Very little attention has been paid to this issue, but, at a rough estimate, domestic cooking and heating with coal in 1979 may have produced a collective effective dose equivalent commitment of 100,000 man-sievert throughout the world.

Little, too, is known of the effect of the fly ash collected by pollution control equipment. In some countries more than a third of it is re-used, mainly in cement and concrete. Some concretes are four-fifths fly ash. It is also used in roadbuilding and to improve farm soil. All of these applications could lead to increased radiation exposures, but very little information has been published in this field.

Geothermal energy is another source of increased radiation. Several countries tap reservoirs of steam and hot water trapped in the earth to generate electricity or to heat buildings; one source has been powering turbines at Larderello in Italy since the turn of the century. Examination of the emissions of this and two very much smaller Italian plants suggests that they produce a collective effective dose equivalent commitment of six

man-sievert per gigawatt year of electricity output – three times the corresponding dose produced by coal fired power stations. Because it accounts for only 0.1% of present world energy production, geothermal energy makes only a tiny contribution to the world's radiation exposure. But it may become very much more important in the future, since many studies suggest that its potential is very great.

Phosphate rock is mined extensively around the world, mainly for use in fertilizers, some 30 million tonnes of which were produced in 1977. Most of the deposits of phosphate ore under exploitation contain high concentrations of uranium. Radon is released during the mining and processing of the ores, and the fertilizers themselves are radioactive and contaminate food. This contamination is normally only slight, but it may be greater if the fertilizer is applied to the soil in liquid form or if phosphate products are fed to animals. Such products are, indeed, widely used in livestock feed supplements and, when fed to dairy cattle, can significantly increase radium levels in milk. All these aspects of the phosphate industry give rise to a collective effective dose equivalent commitment of about 6000 man-sievert, compared to the about 300,000 man-sievert from phosphogypsum from 1977 phosphate production.

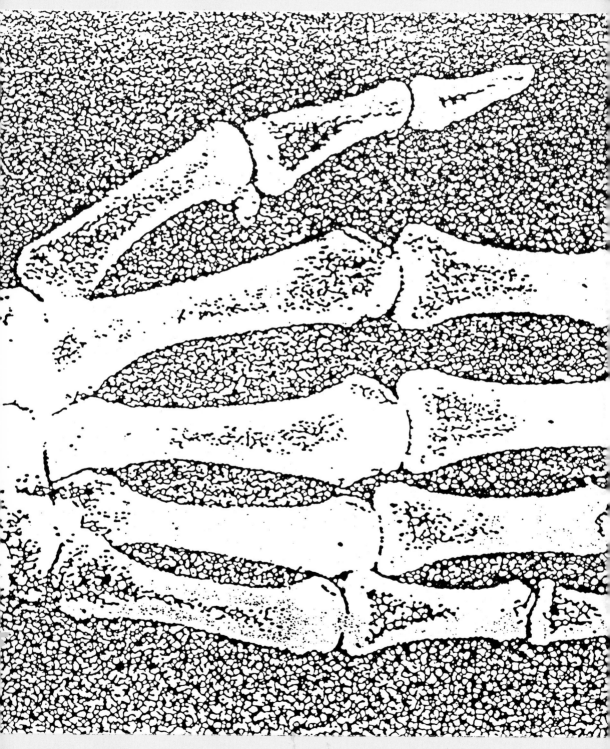

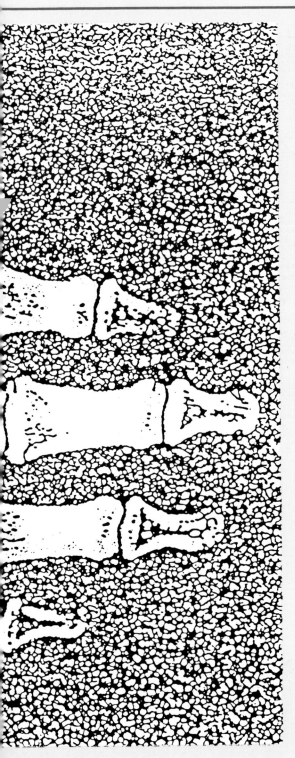

Over the last few decades man has "artificially" produced several hundred radionuclides. And he has learned to use the power of the atom for a wide variety of purposes, from medicine to weapons, from the production of energy to the detection of fires, from illuminating watches to prospecting for minerals. All increase the radiation dose both to individual people and to mankind as a whole.

Individual doses from man-made sources of radiation vary greatly. Most people receive a relatively small amount of artificial radiation; but a few get many thousand times the amount they receive from natural sources.

This variability is generally greater for man-made sources than for natural ones. Most man-made sources, too, can be controlled more readily than most natural ones; though exposure to external irradiation due to fallout from past nuclear explosions, for example, is almost as inescapable and uncontrollable as that due to cosmic rays from beyond the atmosphere or to radiation from out of the earth itself.

Medical sources
At present, medicine is much the greatest source of human exposure from man-made radiation (diagrams 3.1, 4.1). Indeed in many countries it is responsible for nearly all of the dose received from artificial sources.

Radiation is used for both diagnosing and treating disease. The familiar x-ray machines are one of the most useful tools at the service of doctors – and new, sophisticated, diagnostic techniques using radioisotopes are spreading rapidly. Radiation treatment is also, paradoxically, one of the main ways of fighting cancer.

Obviously, individual doses vary enormously – from zero (for someone who has not even had an x-ray examination) to many thousand times the annual average dose from natural radiation (for some patients undergoing treatment for cancer). Yet, there is remarkably little of the reliable and representative information that UNSCEAR needs to calculate doses to the world's population. Not enough is known

4

4.1 TRENDS IN RADIATION SOURCES

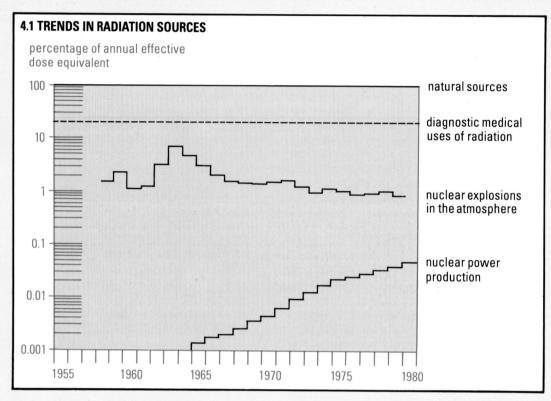

percentage of annual effective
dose equivalent

- natural sources
- diagnostic medical uses of radiation
- nuclear explosions in the atmosphere
- nuclear power production

Trends of different sources of radiation. Annual effective dose equivalents from the various sources are expressed as a percentage of those from natural background radiation. Thus doses from natural sources, of course, remain steady at 100 per cent. Doses from diagnostic medical uses are assumed to remain constant from 1945 to 1980 at 20 per cent of natural background. Doses from atmospheric nuclear explosions, after peaking at around seven per cent in the early 1960s, declined after the Partial Test Ban Treaty to about 0.8 per cent of natural background in 1980. By contrast, doses from nuclear power production have been increasing as the industry expands – from 0.001 per cent of natural background in 1965 to about 0.035 per cent in 1980. Note the logarithmic scale.

about how many people are irradiated each year, or about what doses they receive, or about what parts of their bodies are affected.

In principle, medical radiation is beneficial. But it does seem that people often receive doses that are unnecessarily high. These could be considerably reduced without any loss of efficiency. Since medical radiation accounts for such a high proportion of the exposure to man-made sources, the benefit of such reductions would also be great.

Diagnostic x rays are much the commonest form of medical radiation. Evidence from industrialised countries suggests that annual rates range from 300 to 900 examinations for every 1000 inhabitants – but this excludes dental x-ray examinations and screening by mass radiography. Sparser data from developing countries indicate that their examination rates do not exceed about 100 to 200 per 1000. Indeed about two-thirds of the world's people live in countries where the average frequency of radiological examinations appears to be about a tenth or less of what it is in developed nations.

In most countries about half of all medical x-ray examinations are of the chest. But mass chest x rays are becoming less useful as the incidence of tuberculosis falls. What is more, there is now good evidence that early

4.2 DIAGNOSTIC X RAYS

	Japan	Germany Fed. Rep.	Romania	Chile	Sri Lanka
head	59.1	108.2	37.0	14.7	0.33
chest	640.1	333.9	850.0	45.9	12.9
stomach	285.4	67.8	116.1	17	0.11

number of diagnostic x-ray examinations a year per 1000 inhabitants   O = 10

Annual frequency of diagnostic x-ray examinations of various organs in several countries. (The "stomach" x rays include examinations of the upper gastro-intestinal tract.)

detection of lung cancer by these means does not significantly improve the prospect of survival. The frequency of these examinations has fallen heavily in many industrialised countries including Sweden, the United Kingdom, and the United States. In some other countries, however, about a third of the people are still examined every year.

Over recent years there have been technical improvements which, if applied correctly, should reduce unnecessary doses to patients from x-ray examinations. Disappointingly, studies in Sweden and the United States show that they have generally resulted in little or no reduction in doses.

Doses vary widely from hospital to hospital even within the same country. Several studies in the Federal Republic of Germany, the United Kingdom, and the United States show that the doses delivered by the x-ray beam as it enters the body vary a hundredfold. Meanwhile, another survey has shown that the irradiated area is sometimes twice as large as it should be. Other studies indicate that many facilities produce poor x-ray pictures, and give unnecessary radiation exposure because their equipment performs badly.

Nevertheless, there are cases where radiation exposures have indeed fallen

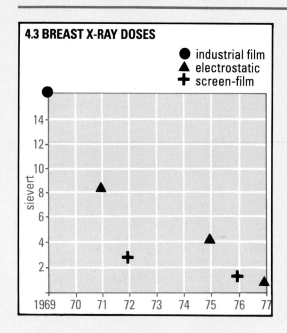

4.3 BREAST X-RAY DOSES

● industrial film
▲ electrostatic
✚ screen-film

Decrease in average doses from best available mammography techniques.

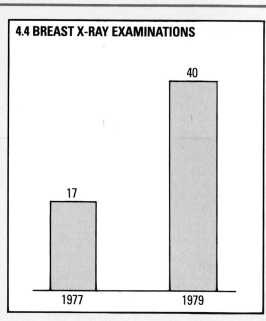

4.4 BREAST X-RAY EXAMINATIONS

Increase in breast x-ray examinations per thousand women in Sweden 1977-9.

because of improved equipment and practice. In other cases, considerable gains in diagnostic efficiency have been made by deliberately increasing doses by moderate amounts. The object must be to keep exposure to the lowest level that is necessary, and UNSCEAR believes there is great potential for significantly reducing doses.

Doses from dental x rays do seem to have come down as a result of technical improvements. This is important, not least because these are the most frequent x-ray examinations in many industrialised countries. Limiting the x-ray beam more tightly, filtering it further to remove unnecessary radiation, using faster films, and adequate shielding all reduce exposure.

Breast examinations have also benefited from reduced doses. Mammographic techniques introduced in the second half of the 1970s generally gave much lower doses than those delivered by earlier equipment (diagram 4.3), and it may be feasible to reduce them further without impairing the

quality of the x-ray picture. This reduction has been matched by an increase in the number of breast examinations undertaken – they more than doubled in both Sweden and the United States between 1977 and 1979 (diagram 4.4).

Another new technique, computed tomography, is considered to be the greatest improvement in the use of radiation for diagnosis since Roentgen's discovery of x rays. Its use, too, is rapidly increasing – it rose a hundredfold in Sweden between 1973 and 1979 (diagram 4.6). A study of kidney examinations showed that the new technique reduced radiation doses to the skin fivefold, to ovaries 25-fold, and to testes 50-fold over conventional techniques (diagram 4.5).

Working out the average doses received by large numbers of people is extremely difficult, partly because data on the frequency of x-ray examinations are so limited – particularly from some of the developing countries. The wide variation in doses from hospital to hospital complicates matters further, because it means that data from

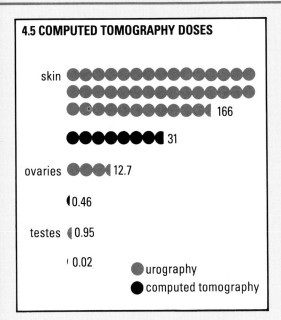

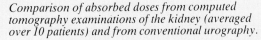

4.5 COMPUTED TOMOGRAPHY DOSES

skin — 166, 31
ovaries — 12.7, 0.46
testes — 0.95, 0.02

● urography
● computed tomography

Comparison of absorbed doses from computed tomography examinations of the kidney (averaged over 10 patients) and from conventional urography.

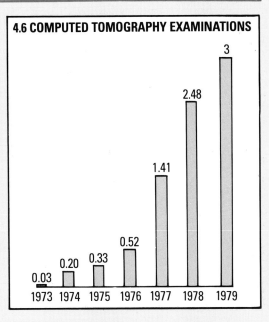

4.6 COMPUTED TOMOGRAPHY EXAMINATIONS

1973: 0.03, 1974: 0.20, 1975: 0.33, 1976: 0.52, 1977: 1.41, 1978: 2.48, 1979: 3

Increase in number of examinations by computed tomography per 1000 people in Sweden, 1973–1979.

one clinic usually cannot be taken to be representative.

Until recently, attempts to assess the average population dose from x-ray examinations have been limited to trying to determine that dose which may entail genetic consequences. It is called the genetically significant dose equivalent, or GSD. Its magnitude depends heavily on two factors. One is whether the patients are likely subsequently to have children; this is strongly influenced by their age. The other is the dose the x rays deliver to the reproductive germ cells. This is linked to the type of examinations carried out; in the United Kingdom the biggest contributors to the GSD in 1977 were examinations of the pelvis and lower back, of the upper femur and hip, of the bladder and urinary tract, and barium enemas.

The GSD in Britain that year was estimated at about 120 microsievert. It was about 150 microsievert in Australia in 1970, the same in Japan in both 1974 and 1979, and about 230 microsievert in the USSR at the

end of the 1970s.

In its latest report, in 1982, UNSCEAR attempted to go further and work out the effective dose equivalent for patients, to assess the potential harm to other tissues in the body beside the reproductive organs. This is difficult to do even in principle because the usual means of calculating this dose is not well suited to medical exposures. There are also technical difficulties. Estimating the effective dose equivalent requires accurate data on how much radiation has been absorbed by up to a dozen different organs or tissues for each examination. The distribution of these doses can differ 1000-fold or more for the same type of x-ray examination – despite technical advances that were actually expected to reduce such variations.

In fact, only two countries, Japan and Poland, could present the Committee with reasonably complete information for calculating these doses – about 600 man-sievert per million people in Poland in 1976, and 1,800 man-sievert per million population in Japan in 1974. In the absence of any other

31

data, UNSCEAR has tentatively assumed that the annual collective effective dose equivalent from x-ray examinations in industrialised countries might be around 1000 man-sievert per million people. In developing countries, of course, the figure is expected to be lower, even though individual doses may be higher.

Radioisotopes are used to explore many bodily processes and to locate tumours. Their use has increased rapidly over the last 30 years, but they are still far less frequent than x-ray examinations. Information is scanty, but what exists suggests that there are only about 10 to 40 examinations per thousand inhabitants in industrialised countries. Estimates of doses are similarly hard to come by; one study in Japan calculated that the annual effective dose equivalent was in the order of 20 man-sievert per person. Collective effective dose equivalents have been found to range from 20 microsievert per million people in Australia to about 150 in the United States.

There are also about 4000 radiotherapy machines in the world. These treat cancer by heavily irradiating malignant tissues to try to kill tumour cells. Yet again, there is very limited information about how much they are used, and about what exposures populations receive. The actual doses given to each patient are high but they are usually given to people with a relatively short life expectancy and little likelihood of having children. They are also given to comparatively few people and so make a very small contribution to the overall dose.

The hundreds of millions of small doses given in x-ray examinations each year far outweigh the relatively few high doses given to cancer patients. The average effective dose equivalent from all medical exposures per inhabitant of industrialised countries may be around one millisievert a year – about half the average dose from natural sources. This estimate conceals wide variations; it may even vary three-fold from country to country. Since developing countries use medical radiation much less, the world average may be about 400 microsievert per person per

year; which would give a total collective effective dose equivalent of about 1,600,000 man-sievert a year.

Nuclear explosions
For the last 40 years everyone has been exposed to radiation from fall-out from nuclear weapons. Virtually none of this comes from the bombs actually dropped on Hiroshima and Nagasaki in 1945; almost all is the result of atmospheric nuclear explosions carried out to test nuclear weapons.

This testing reached two peaks; the first between 1954 and 1958 when the United States, the USSR and the United Kingdom were all exploding devices; the second, and greater, in 1961 and 1962 when the United States and the USSR were the main contributors. During the first period, United States' tests dominated – during the second, tests by the Soviet Union (diagram 4.7).

In 1963 these three countries signed the Partial Test Ban Treaty, undertaking not to test nuclear weapons in the atmosphere, oceans, and outer space. Since then France and China have conducted a series of much smaller atmospheric tests of declining frequency – the latest in 1980. Underground tests are still being carried out, but they generally give rise to virtually no fall-out.

Some of the radioactive debris from atmospheric tests lands relatively close by. Some stays in the troposphere, the lowest layer of the atmosphere, and is carried by the wind around the world at much the same latitude; as it travels it gradually falls to earth, remaining, on average, about a month in the air (diagram 4.8, overleaf). But most is pushed into the stratosphere, the next layer of the atmosphere (from about 10 to 50 kilometres up), where it stays for many months, and whence it slowly descends all over the earth.

These various types of fall-out contain several hundred different radionuclides, but only a few contribute much to human exposure as most are produced in very small

Atmospheric nuclear tests and their yields.

4.7 ATMOSPHERIC NUCLEAR TESTS

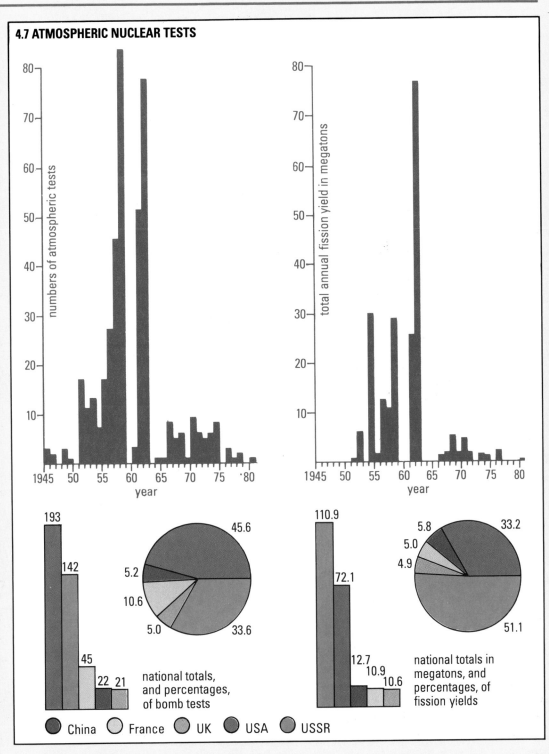

national totals, and percentages, of bomb tests

national totals in megatons, and percentages, of fission yields

China   France   UK   USA   USSR

amounts or decay quickly. Only four contribute more than one per cent to the collective effective dose equivalent commitment of the world population from nuclear explosions. These are, in declining order of importance, carbon-14, caesium-137, zirconium-95 and strontium-90.

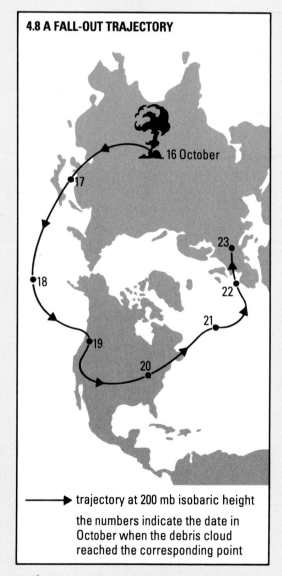

4.8 A FALL-OUT TRAJECTORY

16 October

17

23

18

22

21

19

20

→ trajectory at 200 mb isobaric height

the numbers indicate the date in October when the debris cloud reached the corresponding point

Tropospheric fall-out from an atmospheric nuclear explosion on 16 October, 1980. Only one of several trajectories traced at different isobaric heights is shown.

The doses from these, and other radionuclides, are delivered over different periods, because they decay at different rates. Thus zirconium-95, which has a half-life of 64 days, has already delivered practically all its dose. Caesium-137 and strontium-90 both have half-lives of about 30 years, and so will have deposited most of their doses by the end of the century. Only carbon-14, with its 5,730 year half-life, will stay active into the far future, though at low dose rates; by the year 2000 it will have delivered only seven per cent of its eventual contribution.

Annual doses have closely followed testing, peaking in 1958–1960 and, especially, in 1963–1964 (diagrams 4.9, 4.10 and 4.11, overleaf). In 1963 the average annual collective doses amounted to about seven per cent of the equivalent exposure to natural radiation; this decreased to two per cent by 1966, and one per cent by the early 1980s. If no more atmospheric tests take place, future annual doses will get smaller and smaller.

These averages do conceal considerable variations. The northern hemisphere, where most of the testing has taken place, has also received most of the fall-out. Reindeer herders in the extreme north accumulate doses of caesium-137 of between a hundred and a thousand times normal levels, just as they receive more natural radiation; the caesium concentrates in the lichen-reindeer food chain. Unfortunately, some people in the vicinity of test sites, such as some inhabitants of the Marshall Islands, and a boatload of Japanese fishermen who happened to pass nearby, have received high doses.

The total collective effective dose equivalent commitment from atmospheric nuclear explosions conducted so far amounts to 30,000,000 man-sievert. Only twelve per cent of this had been delivered by 1980, the rest will reach man over millions of years.

Nuclear power
The production of nuclear power is much the most controversial of all the man-made sources of radiation – yet at present it makes

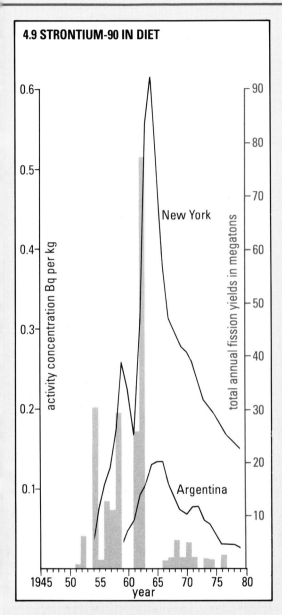

4.9 STRONTIUM-90 IN DIET

activity concentration Bq per kg

total annual fission yields in megatons

New York

Argentina

1945 50 55 60 65 70 75 80
year

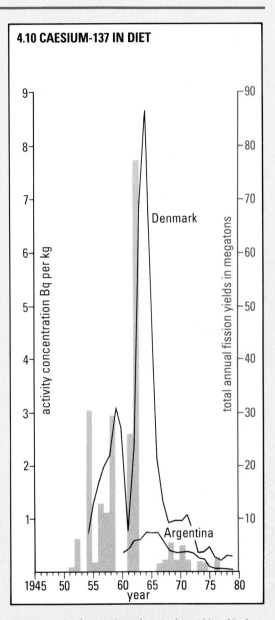

4.10 CAESIUM-137 IN DIET

activity concentration Bq per kg

total annual fission yields in megatons

Denmark

Argentina

1945 50 55 60 65 70 75 80
year

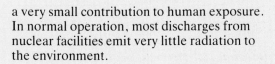

Levels of strontium-90 and caesium-137 in the total diet, compared with the annual yield from atmospheric nuclear tests. Note the much higher exposures in the northern hemisphere (New York and Denmark) than in the southern hemisphere (Argentina).

a very small contribution to human exposure. In normal operation, most discharges from nuclear facilities emit very little radiation to the environment.

By the end of 1984 there were 345 nuclear power reactors in operation in 26 countries, world-wide. They produced 13 per cent of the world's electricity from a total generating

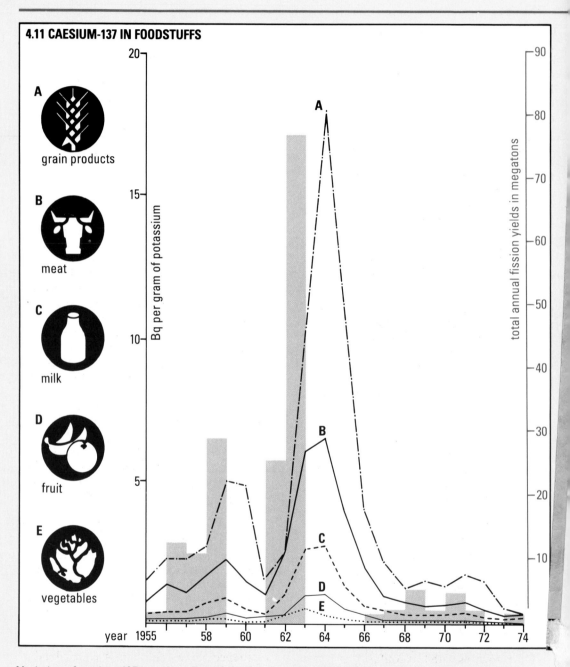

4.11 CAESIUM-137 IN FOODSTUFFS

A
grain products

B
meat

C
milk

D
fruit

E
vegetables

Variation of caesium-137 in various Danish foods. The yearly yield from atmospheric nuclear tests is shown for comparison.

capacity of 220 gigawatts (diagram 4.12. overleaf). Capacity had then doubled in little over five years – but future growth rates are unclear. Predictions of nuclear capacity at the end of the century have been steadily falling over the last years as the projected expansion

36

of nuclear power slows under the impact of economic recession, energy conservation, and public opposition. The latest forecast by the International Atomic Energy Agency, in 1983, projects a world-wide capacity for the year 2000 of between 720 and 950 gigawatts of electricity.

These power stations are just part of the nuclear fuel cycle. This starts with the mining and milling of uranium ore and proceeds to the making of nuclear fuel. After use in power stations, the irradiated fuel is sometimes "reprocessed" to recover uranium and plutonium. Eventually the cycle will end with the disposal of nuclear wastes (diagram 4.14, see page 42).

At each stage in this cycle radioactive materials are released. UNSCEAR has set out to evaluate the doses delivered to the public at each part of the cycle both in the short term and over many hundreds of years. This is a complicated and difficult undertaking. For a start, emissions vary widely, even from similar installations; the levels of radioactive gases given off by boiling water reactors (BWRs), for example, can vary more than a millionfold from plant to plant and year to year.

The doses also vary over space and time. Generally speaking the further people live from a particular nuclear installation the less radiation they will receive from it; and whereas some installations are in remote areas, others are near centres of population. These installations emit a variety of radionuclides which decay at different rates; most are of only local importance because they decay rapidly; some live long enough to spread right around the world; and some remain in the environment virtually for ever. Different radionuclides also behave differently in the environment; some spread quickly, others move very little.

To get to grips with this confusing situation, UNSCEAR has developed hypothetical model installations for each step of the fuel cycle, designed to be typical facilities, in typical geographical areas, surrounded by typical population densities. It has also studied information on discharges from the world's nuclear plants and produced average releases for each gigawatt year of electricity generated. These generalisations give an idea of the overall impact of the nuclear power programme, but, obviously, cannot be applied indiscriminately to any individual plant. They must be treated with the utmost caution; they cannot be taken at face value, and are subject to a large number of assumptions spelled out in the UNSCEAR reports.

About half the world's uranium ore comes from open-cast mines, and half from underground ones. It is then taken to mills, usually nearby, for processing. Both mines and mills give off radioactive discharges to the environment. The mines account for nearly all the combined dose from the two operations in the short term. But the mills are responsible for a much greater long-term problem; they produce large amounts of waste, or "tailings" – 120 million tonnes are already stored at active mill sites, mainly in North America. If current trends continue there will be 500 million tonnes by the end of the century.

These wastes remain radioactive for millions of years after mills cease operation, thus providing potentially the greatest long term contribution to human exposure from nuclear power. But this contribution could be reduced greatly, at least in the short term, if the tailings were covered with asphalt or polyvinylchloride. Such covers would, of course, have to be regularly replaced.

After leaving the mills, the uranium is turned into fuel by further processing and purification and, usually, by passing through an enrichment plant. These processes give rise to both airborne and liquid discharges but the doses are very much smaller than from other parts of the fuel cycle.

The fuel is now ready to be used in reactors to produce power. There are five main kinds of reactor in operation; pressurised water reactors and boiling water reactors, which were originally developed in the United States and are now the commonest types in the world; gas-cooled reactors, developed, and predominantly used, by the United

Kingdom and France; heavy water reactors, largely confined to Canada; and light-water cooled graphite moderated reactors, which are in operation only in the USSR. Besides these, there are also four fast breeder reactors, which are envisaged as the next generation of nuclear power plants in Europe and the USSR.

The quantities of different types of radioactive materials released from these

4.12 THE WORLD OF NUCLEAR POWER

Countries with nuclear power reactors in operation at the end of 1984

source: International Atomic Energy Agency Annual Report for 1984.

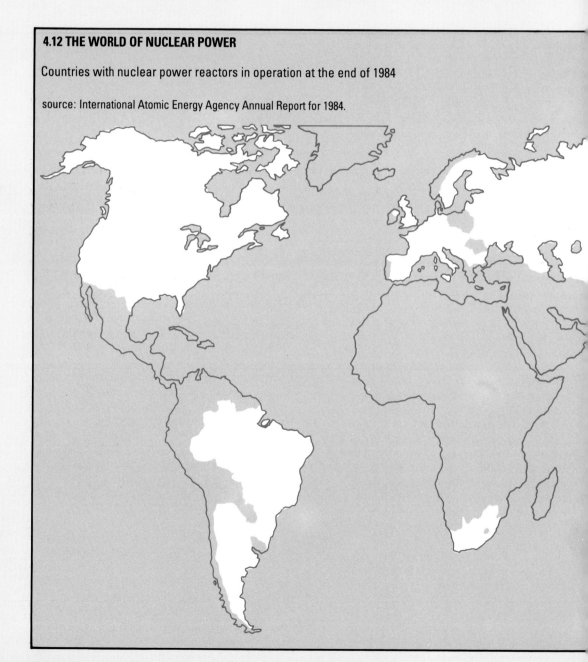

reactors vary widely, not only from type to type, not only between different designs within these types, but even between reactors of the same design. They also vary from year to year for the same reactor, partly because

the amount of maintenance work (which gives rise to the greatest routine discharges) differs each year.

Recently, releases from reactors have tended to decrease despite the increasing

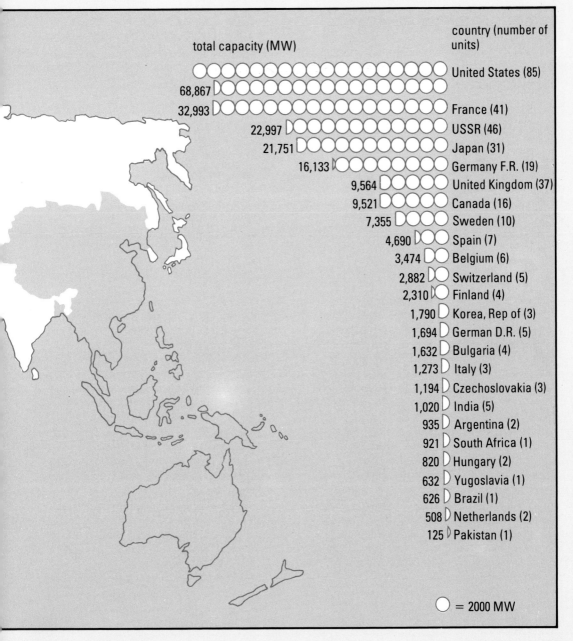

total capacity (MW)

country (number of units)

68,867	United States (85)
32,993	France (41)
22,997	USSR (46)
21,751	Japan (31)
16,133	Germany F.R. (19)
9,564	United Kingdom (37)
9,521	Canada (16)
7,355	Sweden (10)
4,690	Spain (7)
3,474	Belgium (6)
2,882	Switzerland (5)
2,310	Finland (4)
1,790	Korea, Rep of (3)
1,694	German D.R. (5)
1,632	Bulgaria (4)
1,273	Italy (3)
1,194	Czechoslovakia (3)
1,020	India (5)
935	Argentina (2)
921	South Africa (1)
820	Hungary (2)
632	Yugoslavia (1)
626	Brazil (1)
508	Netherlands (2)
125	Pakistan (1)

◯ = 2000 MW

4.13 REPROCESSING PLANTS

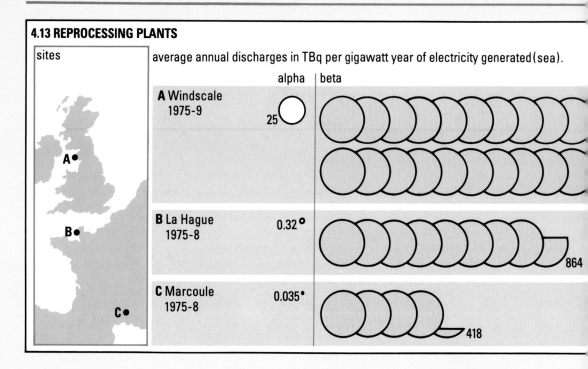

sites

average annual discharges in TBq per gigawatt year of electricity generated (sea).

alpha | beta

A Windscale 1975-9   25

B La Hague 1975-8   0.32   864

C Marcoule 1975-8   0.035   418

electrical output of plants. This is partly due to technological improvements, and partly due to stricter radiation protection measures.

After use in power stations, less than a tenth of the world's irradiated fuel is reprocessed to remove uranium and plutonium for re-use. Only three commercial reprocessing plants are now known to be in operation; at Marcoule and La Hague in France, and at Windscale (Sellafield) in the United Kingdom. Marcoule, which is particularly tightly controlled because it discharges to the River Rhone, is by far the cleanest. Of the other two plants, both of which discharge to the sea, Windscale is very much the more polluting, although a large part of the radioactive material it releases does not come from the reprocessing itself, but from the corrosion of fuel cans stored while awaiting treatment.

Between 1975 and 1979 Windscale's discharges to the sea were responsible for more than three and a half times more beta

activity – and 75 times more alpha activity – for each gigawatt year of electricity produced than were La Hague's during much the same period (diagram 4.13).

Windscale has since been greatly improving its discharges, but it is still more polluting for each unit of fuel processed than La Hague. Hopefully, releases from future reprocessing plants will be very much lower than from either. Notional designs exist which provide for very low discharges to water, and UNSCEAR has based its model facility on the implied releases from a new plant planned for Windscale.

No disposal of highly radioactive wastes from nuclear power production – the last stage in the fuel cycle – has yet taken place. National authorities are storing them; and some have been researching ways of solidifying them and disposing of them in stable geological formations on land, or on, or under, the seabed. Once disposal has taken place, virtually none of the

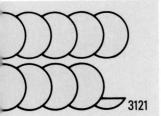

3121

average collective dose to workers in man-Gy per gigawatt year of electricity produced.

A Windscale: average 1971–1978   **B** La Hague: average 1972–1976

18   6

average collective effective dose equivalent commitment via sea in man-Sv per gigawatt year of electricity produced, 1975–1979.

A Windscale   **B** La Hague

124   53

activity in the wastes is supposed to reach man in the foreseeable future. UNSCEAR has made no assessment of the dose commitments from these wastes, though an attempt was made by the International Fuel Cycle Evaluation, in 1979, to evaluate the fate of materials disposed underground. This estimated that it would be between a hundred thousand and a million years before any significant amounts of radioactive materials reached the biosphere.

In all, UNSCEAR estimates, the operation of the fuel cycle contributes a short-term collective dose equivalent commitment of about 5.5 man-sievert for every gigawatt year of electricity produced by the world's nuclear reactors (diagram 4.14, overleaf). Mining contributes 0.5 man-sievert; milling 0.04 man-sievert; and fuel fabrication a mere 0.002 man-sievert. Nuclear reactors are responsible for the bulk of the dose, contributing about four man-sievert, and reprocessing accounts for another one man-

sievert. The figure for reprocessing, as indicated above, is compiled from estimates of the notional effluents from future plants. Present-day facilities actually give rise to doses ten to twenty times higher – but as they treat less than ten per cent of the world's fuel, the overall contribution is much the same.

Ninety per cent of this short-term dose is delivered within a year of the discharge of the radioactive materials – 98 per cent within five years. Almost all of it is received by the local and regional populations, people living within a few thousand kilometres of the installations.

The fuel cycle operations also give off much longer-lived radionuclides which become distributed around the globe. UNSCEAR estimates the collective effective dose equivalent commitment from this source as 670 man-sievert for every gigawatt year of electricity produced, of which less than three per cent will be delivered during the first 500 years. These longer-lived radionuclides

4.14 THE FUEL CYCLE

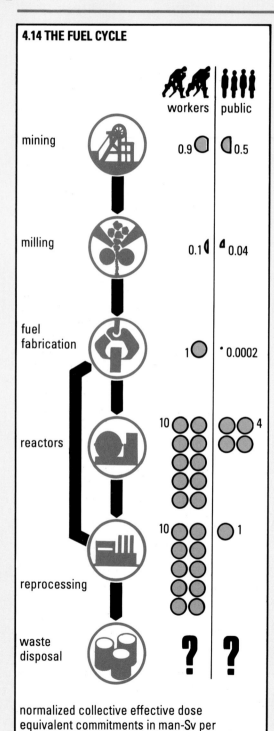

	workers	public
mining	0.9	0.5
milling	0.1	0.04
fuel fabrication	1	0.0002
reactors	10	4
reprocessing	10	1
waste disposal	?	?

normalized collective effective dose equivalent commitments in man-Sv per gigawatt year of electricity generated

deliver much the same annual average doses to the world population as the short-term radiation does to the regional and local inhabitants, but over a much longer period – 90 per cent of the dose is delivered between a thousand and a hundred million years after release. People living near a plant, therefore, will normally receive all the short term dose and a small part of the long term one.

These figures do not include doses contributed by mill tailings and nuclear waste. The effect of waste disposal is thought to be negligible over the next few thousand years, and to contribute only 0.1 to one per cent of the total dose commitment from nuclear power thereafter. But the tailings will, without dispute, contribute a considerable problem if not properly covered. When these two sources are taken into account, the total collective effective dose equivalent commitment from long-lived radionuclides reaches about 4,000 man-sievert for every gigawatt year of electricity produced. But these estimates are necessarily very uncertain. There are great difficulties in making them, not least in predicting future waste management techniques and practices, population sizes and habits, since most of the dose is not delivered for 10,000 years. So UNSCEAR warns against using these figures in decision-making and suggests that little significance should be attributed to them.

The annual collective effective dose caused by the nuclear fuel cycle in 1980 was about 500 man-sievert. By the year 2000 this should rise to 10,000 man-sievert and, by 2100 AD, to 200,000 man-sievert. This is on the pessimistic assumption that no technical improvements are made and that the present levels of discharges continue. Even so, average doses remain a small fraction of the exposure received from natural radiation – rising to one per cent in the year 2100.

The nuclear fuel cycle and the environmental and occupational doses arising from its various stages. The doses are expressed as normalized collective effective dose equivalent commitments in man-sievert per gigawatt year of electricity generated.

People living near nuclear installations do, of course, receive much higher doses than the average. Even so, typical doses around nuclear reactors at present range between a fraction of one per cent and a few per cent of doses from natural sources. Meanwhile, the dose received even by people most at risk around Windscale from the 1979 releases of caesium-137 was probably under a quarter of what they received from natural radiation in the same year.

All the above figures, of course, assume that the nuclear plants operate normally; for very much larger quantities of nuclear materials may be released in accidents. In its last report UNSCEAR made a first attempt to evaluate such doses by examining the accidents at Three Mile Island in 1979 and Windscale in 1957. Releases from Three Mile Island were small; but the Windscale accident, it has been estimated, produced a collective effective dose equivalent commitment of 1,300 man-sievert. The committee found it impossible, however, on the basis of those two accidents, to estimate the general contribution of accidental releases to doses either in retrospect or in the future.

Occupational exposures

The people who get the largest doses of radiation from the nuclear power industry are those who work in it. As in almost all industries, the biggest exposures are occupational.

Difficulties bedevil attempts to assess occupational doses; conditions vary widely and there is not enough information. Exposures inside nuclear facilities differ just as emissions do; and the various devices used to monitor radiation doses are designed to ensure that workers are not exposed to undue levels and seldom provide the kind of information required for detailed dose assessments.

Estimates for uranium mines and mills suggest that their workers receive averages of one man-sievert of radiation for every gigawatt year of electrical energy eventually generated from their production. Again,

mines are responsible for about 90 per cent of this dose and, naturally, underground mines subject the people who work in them to greater doses than do open-cast ones. Fuel manufacturing facilities also probably produce a collective dose equivalent of one man-sievert per gigawatt year (diagram 4.14).

These figures conceal a wide variety of doses, and such variations are even more marked in nuclear reactors. Measurements at pressurised water reactors, for example, show that collective doses per gigawatt year of electricity varied a hundredfold in 1979. By and large newer power stations produced lower doses than older ones. On average, most types of reactors seem to produce annual collective effective dose equivalents of 10 man-sievert per gigawatt year.

Different jobs give workers different doses (diagram 4.15, overleaf). Maintenance work, whether routine servicing or unscheduled repairs, accounts for much the greatest part of the collective dose – about 70 per cent in United States reactors. Sometimes workers are brought in on contract to do this particularly dirty work. In the United States such contract workers receive half the total collective dose.

Large numbers of workers get significant doses at both the Windscale and La Hague reprocessing plants. Again, there is a difference between the two facilities; during the 1970s, Windscale delivered an average annual collective dose per gigawatt year of 18 man-sievert, three times the La Hague level (diagram 4.13). But new reprocessing plants are likely to give much lower doses. UNSCEAR estimates that 10 man-sievert per gigawatt year may be a realistic future global figure.

Workers engaged in nuclear research and development receive doses that vary particularly widely between plants and countries. Collective doses per unit of electrical energy produced differ tenfold from country to country; they are, for example, low in Japan and Switzerland, and high in the United Kingdom. A realistic global figure might be five man-sievert per gigawatt year.

These estimates add up to a rough total

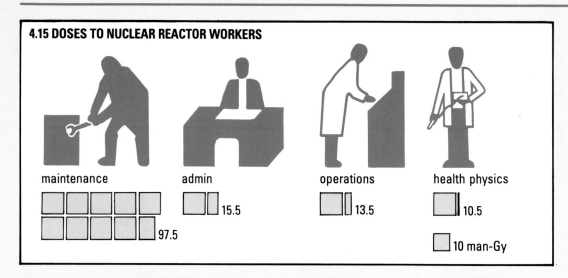

4.15 DOSES TO NUCLEAR REACTOR WORKERS

maintenance 97.5

admin 15.5

operations 13.5

health physics 10.5

10 man-Gy

Different doses from different jobs at nuclear reactors. The diagram shows average annual collective doses (in man-Gy) received between 1977 and 1979 by workers in PWRs and BWRs in the United States.

annual collective dose equivalent of rather less than 30 man-sievert for every gigawatt year of electricity generated – or a total of 2000 man-sievert in 1979. This is about 0.03% of the corresponding dose from natural sources.

This figure, which extends the occupational doses over the entire population, obscures the fact that radiation workers receive greater doses from their work than from natural sources. Underground uranium miners have traditionally received the highest average doses, over six times the average from natural sources, but Windscale workers now run them close. Open cast miners, and workers at La Hague, and at PWR, BWR and HWR power stations receive an occupational average dose of about twice what people normally get from natural sources. Only workers in gas-cooled reactors and fuel fabrication plants receive average additional doses of about the same magnitude as the average from natural radiation. And these average occupational doses conceal wide individual variations.

Of course, it is not only workers in the nuclear industry who receive occupational doses of man-made radiation. Medical and general industrial workers are also exposed. Exposure to medical personnel (diagram 4.16) involves relatively low average doses to large numbers of workers (there are at least 100,000 in the United States alone; more in Japan and the Federal Republic of Germany). Annual average doses for dentists in different countries are even lower. Overall, it is estimated that the exposure of medical personnel engaged in radiological work contributed about one man-sievert per million people to the collective equivalent dose in countries with a high standard of medical care.

The use of radiation in general industry may produce an annual collective dose of about another 0.5 man-sievert per million population in industrialised countries. Many thousands of workers seem to be exposed, but little is known about them. The relatively small numbers who use radioactive materials for making products luminous receive high annual average doses.

Industrial radiographers use radiation under what are often rather primitive conditions on building sites and the like. They are widely believed to be among the most highly exposed of all workers, though hard

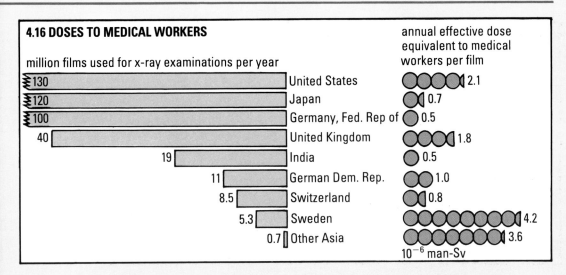

4.16 DOSES TO MEDICAL WORKERS

annual effective dose
equivalent to medical
workers per film

million films used for x-ray examinations per year

130	United States	2.1
120	Japan	0.7
100	Germany, Fed. Rep of	0.5
40	United Kingdom	1.8
19	India	0.5
11	German Dem. Rep.	1.0
8.5	Switzerland	0.8
5.3	Sweden	4.2
0.7	Other Asia	3.6

10^{-6} man-Sv

*Medical occupational doses and medical practice.
The diagram gives an indication of the efficiency of
the precautions against such occupational exposures
in different countries by showing the average
effective dose equivalent to medical workers per
film.*

evidence is not easy to come by. Certainly,
they are the most likely to suffer from
accidental overexposures to radiation.

Some workers are exposed to higher doses
of **natural** radiation than usual by their work.
Aircrew form the largest group: the altitude
at which they work increases their exposure
to cosmic rays. Some 70,000 aircrew in the
United States, and 20,000 in the United
Kingdom, receive an average of between one
and two millisievert extra a year.

Far beneath them, coal and metal miners
also receive enhanced doses. These are highly
variable, but – in some forms of underground
mining, other than coal mining – can rival the
higher exposures found in uranium mines.
Workers at radon spas, where people go for
supposedly beneficial treatment, can receive
very high doses, sometimes exceeding 300
millisievert a year – six times the
internationally recommended limit for
nuclear workers.

Miscellaneous sources
Finally, some common consumer products
contain materials which expose a generally
unaware public to radiation.

Luminous watches and clocks provide

much the biggest world-wide dose. They have
an annual impact four times as great as
environmental releases from nuclear energy,
and give rise to the same collective effective
dose equivalent as from air travel or from
occupational exposures in the nuclear
industry – 2000 man-sievert (diagram 4.18,
overleaf).

Watches used to be luminised with radium.
This exposes the whole body of the owner to
penetrating radiation – though the dose is
10,000 times greater one centimetre from the
dial than it is one metre away. It is now
tending to be replaced by tritium or
prometheum-147, which give much smaller
doses. Yet by the end of the 1970s, 800,000
watches containing radium were still in use in
the United Kingdom. International standards
for radium in watches were published in 1967,
but many watches still in use may predate
them. Radionuclides are also used to
illuminate exit signs, compasses, gun sights,
telephone dials, and many other devices.

Anti-static brushes using alpha particles are
sold in the United States to remove dust from
gramophone records and photographic
equipment. In 1975 the United Kingdom's
National Radiological Protection Board

found they could be dangerous in certain circumstances.

Many smoke detectors also use alpha radiation. More than 26 million of them, incorporating americium-241, had been installed in the United States by the end of the 1980s, but they only emit tiny doses in normal use. Radionuclides are also used in starters for fluorescent tube lights and some electrical appliances. Nearly a hundred million such products were in use in West Germany alone in the mid 1970s. Unless they are broken, they do not cause significant doses.

Thorium is used in some specially thin optical lenses, and can deliver substantial doses to the lens of the eye. Uranium is commonly used in false teeth to make them shine, and can irradiate the tissues of the mouth. The U.K. National Radiological Protection Board has recommended that its use is discontinued; and the United States and the Federal Republic of Germany, which make most dental porcelain, limit concentrations. Both these uses are purely for aesthetic reasons, and so the resulting exposures are entirely unjustified.

X rays are produced inside colour televisions, but modern sets emit only negligible amounts as long as they are used normally and serviced appropriately. X-ray machines used for screening baggage give air travellers only minute doses each trip. More seriously, disturbing surveys in the United States and Canada in the early 1970s showed that many secondary schools were using x-ray tubes that could emit high doses – and that most teachers demonstrating them had little or no knowledge of radiation protection.

*Comparison of doses from different man-made sources of radiation. Some of the doses assessed in the last two chapters are displayed in these two diagrams to aid comparisons. Diagram 4.17 below looks at a number of annual collective effective dose **commitments** for specific years. Table 4.18, right,* *makes a similar comparison for the **annual** collective effective dose equivalents from a number of other sources. It also shows how this dose from nuclear power production is expected to grow with the expansion of the industry over the next 200 years or so.*

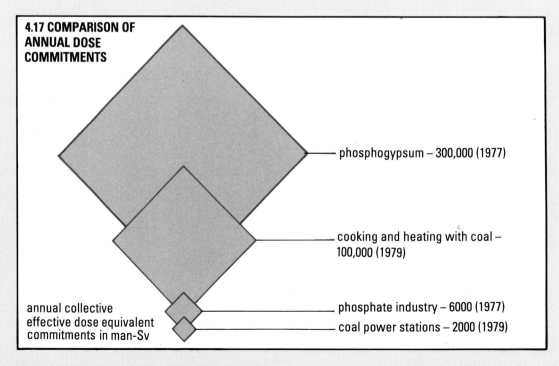

4.17 COMPARISON OF ANNUAL DOSE COMMITMENTS

phosphogypsum – 300,000 (1977)

cooking and heating with coal – 100,000 (1979)

phosphate industry – 6000 (1977)

coal power stations – 2000 (1979)

annual collective effective dose equivalent commitments in man-Sv

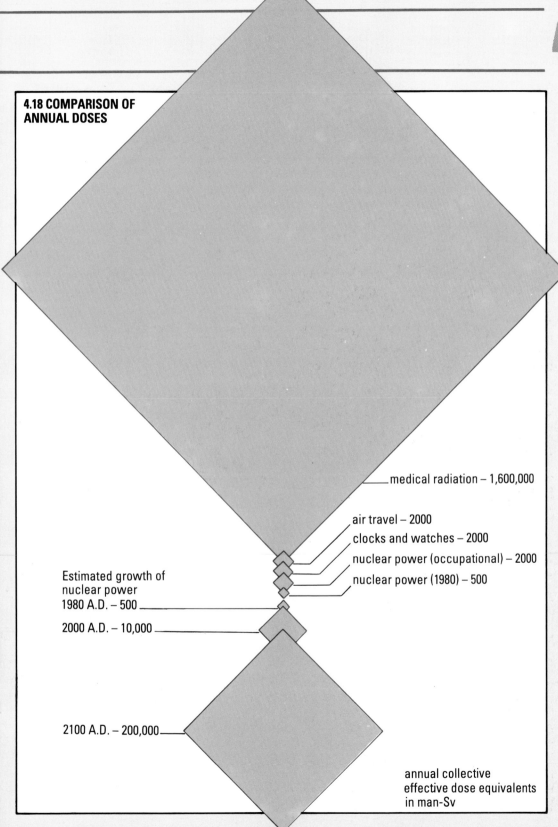

4.18 COMPARISON OF ANNUAL DOSES

medical radiation – 1,600,000

air travel – 2000

clocks and watches – 2000

nuclear power (occupational) – 2000

nuclear power (1980) – 500

Estimated growth of
nuclear power
1980 A.D. – 500

2000 A.D. – 10,000

2100 A.D. – 200,000

annual collective
effective dose equivalents
in man-Sv

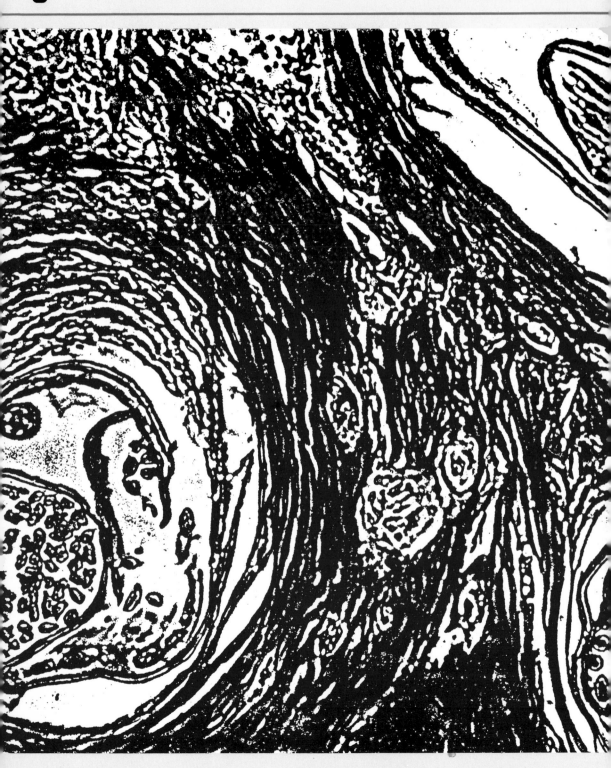

Radiation, by its very nature, is harmful to life. At low doses, it can start off only partially understood chains of events which lead to cancer or genetic damage (diagram 5.1, overleaf). At high doses, it can kill cells, damage organs, and cause rapid death.

The damage done by high doses normally becomes evident within hours or days. Cancers, however, take many years – usually decades – to emerge. And, by definition, the hereditary malformations and diseases caused by genetic damage take generations to show; it is the children, grandchildren, or remoter descendants of the people originally irradiated who will be affected.

Whereas it is usually quite easy to identify the early, acute effects of high doses, it is almost always extremely hard to pin down these "late" effects from low ones. This is partly because you have to wait much longer for them to become evident. Even then, it is hard to apportion blame because both cancer and genetic damage are not specific to radiation but have many other causes.

Radiation doses have to reach a certain level to produce acute injury – but not to cause cancer or genetic damage. In theory, at least, just the smallest dose can be sufficient. So, no level of exposure to radiation can be described as safe. Yet, at the same time, no level is uniformly dangerous. Even at quite high doses not everyone is affected; the body's repair mechanisms usually offset the damage that is done. Similarly, someone exposed to a dose of radiation is by no means fated to develop cancer or sustain genetic damage; but he is at greater **risk** than if he had not been irradiated. And the size of the risk will increase with the size of the dose.

UNSCEAR tries to work out, as reliably as possible, just what extra risks people face from different doses of radiation. Probably more research has been carried out on the effects of radiation than on any similar hazard. But the longer-term the effect, and the lower the dose, the less immediately useful information there is.

Acute effects
In its most recent report, UNSCEAR carried

49

out an extensive review, for the first time in 20 years, of the acute effects caused by high doses of radiation. Generally speaking, the damage appears only after a minimum, or "threshold" dose.

A vast amount of information has accumulated from the use of radiotherapy to treat cancer. Over the years, the medical profession has learned a great deal about how human tissues respond to radiation. Different parts of the body vary enormously in their reaction (diagram 5.3, overleaf). And the size of the dose needed to cause damage depends on whether it is given all at once or in instalments. Most organs can repair radiation damage to some extent and so can tolerate a series of smaller doses better than the same total dose given in a single exposure.

Of course, if the dose is high enough, the person irradiated will die. Generally speaking, very high doses, of 100 gray or so, damage the central nervous system so badly that death may occur within hours or days (diagram 5.2). At doses of 10 to 50 gray to the whole body, the victim may escape this fate only to die from gastrointestinal damage between one and two weeks later. Lower doses still may avoid gastrointestinal injury – or permit recovery from it – but still cause death after a month or two, mainly from damage to the red bone marrow – the tissue that forms blood; a whole-body dose of about three to five gray will kill half the people who receive it. So the higher doses merely hasten the time of death. Of course, it is very often a combination of these conditions that proves fatal. The whole area is an important subject for study because the information is needed

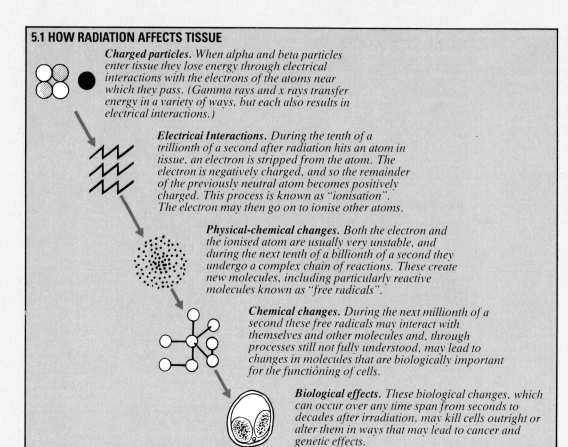

5.1 HOW RADIATION AFFECTS TISSUE

Charged particles. When alpha and beta particles enter tissue they lose energy through electrical interactions with the electrons of the atoms near which they pass. (Gamma rays and x rays transfer energy in a variety of ways, but each also results in electrical interactions.)

Electrical Interactions. During the tenth of a trillionth of a second after radiation hits an atom in tissue, an electron is stripped from the atom. The electron is negatively charged, and so the remainder of the previously neutral atom becomes positively charged. This process is known as "ionisation". The electron may then go on to ionise other atoms.

Physical-chemical changes. Both the electron and the ionised atom are usually very unstable, and during the next tenth of a billionth of a second they undergo a complex chain of reactions. These create new molecules, including particularly reactive molecules known as "free radicals".

Chemical changes. During the next millionth of a second these free radicals may interact with themselves and other molecules and, through processes still not fully understood, may lead to changes in molecules that are biologically important for the functioning of cells.

Biological effects. These biological changes, which can occur over any time span from seconds to decades after irradiation, may kill cells outright or alter them in ways that may lead to cancer and genetic effects.

to predict the effects of nuclear war and of high radiation doses from nuclear accidents.

The red bone marrow and the rest of the blood-forming system are among the most sensitive parts of the body, and are affected by as little as 0.5 to one gray. Fortunately, they also have a remarkable capacity for regeneration and, if the dose is not so great as to overwhelm them, can completely recover. If only part of the body is irradiated, enough bone marrow will normally survive unimpaired to replace what is damaged.

Reproductive organs and eyes are also particularly sensitive. Single doses of as little as 0.1 gray to the testes have made men temporarily sterile, and doses over two gray can cause permanent sterility. The testes seem to be unique in that doses given in instalments cause more, not less, damage than the same exposure given all at once – and many years can pass after severely damaging doses before the testes produce sperm fully again. The ovary is rather less sensitive, at least in adult women. But single doses over three gray will still cause sterility, though higher doses can be given in instalments without impairing fertility.

The lens is the part of the eye most vulnerable to radiation. As its cells die, they become opaque, and as the opacities grow they can lead to cataracts, and to total blindness. The higher the dose, the greater the loss of vision. Single doses of two gray or less can create opacities, and more serious progressive cataracts occur with doses of five gray. Even occupational exposure has been shown to affect the eye; doses of 0.5 to two gray over 10 to 20 years increase the density

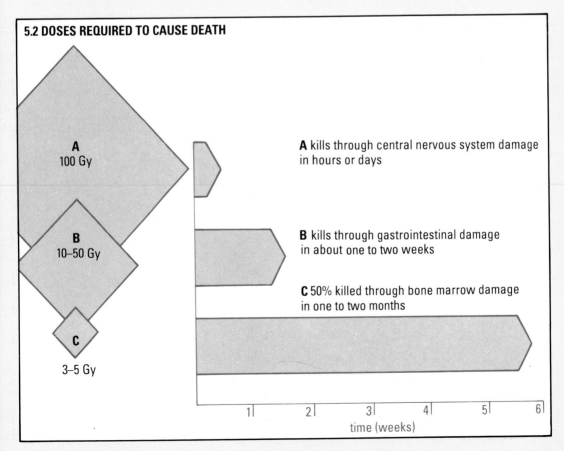

5.2 DOSES REQUIRED TO CAUSE DEATH

A 100 Gy

B 10–50 Gy

C 3–5 Gy

A kills through central nervous system damage in hours or days

B kills through gastrointestinal damage in about one to two weeks

C 50% killed through bone marrow damage in one to two months

time (weeks)

5.3 "ACCEPTABLE" DOSES IN RADIOTHERAPY

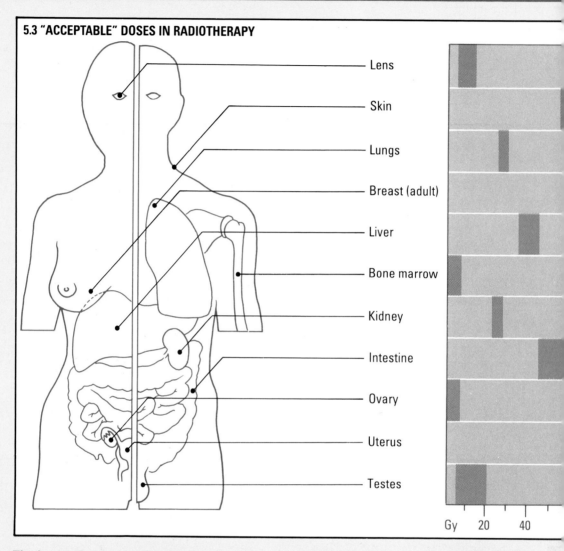

Lens
Skin
Lungs
Breast (adult)
Liver
Bone marrow
Kidney
Intestine
Ovary
Uterus
Testes

Gy 20 40

The doses in this diagram, modified from P. Rubin and G.W. Casarett, in Clinical Radiation Pathology *(Saunders, Philadelphia, 1968), can be acceptably given to patients in five instalments per week. The definition of "acceptable" is the authors',* *not UNSCEAR's, though a fuller table of their conclusions is printed in UNSCEAR's 1982 report. The diagram provides a rough illustration of the differing sensitivities of different organs and tissues.*

and opacity of the lens.

Children are also particularly sensitive. Quite small doses to their cartilage can slow or halt the growth of their bones, and lead to deformity. The younger the child, the more severe is the stunting. Total doses of 10 gray given in daily instalments over a few weeks are sufficient to cause some deformity. Indeed, there may be no threshold at all for this particular effect. Similarly, irradiation of children's brains during radiotherapy has caused changes in character, loss of memory, and even, in very young children, dementia and idiocy. Adult bones and brains can

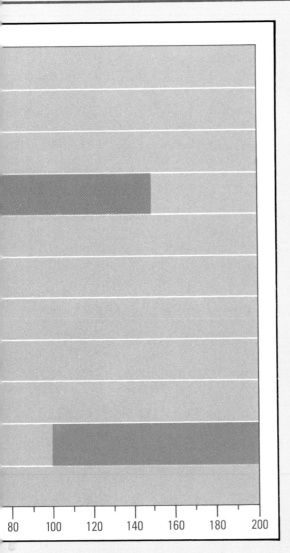

80	100	120	140	160	180	200

distressing – the number of women at that particular stage of pregnancy at any one time forms only a small proportion of the population. This is, however, the most important known effect of irradiating unborn children – though many other effects including malformations, growth retardation, and death have been found in animal foetuses and embryos.

Most adult tissues are relatively robust in their response to radiation. The kidney will take about 23 gray over five weeks, without significant signs of damage, the liver at least 40 gray over a month, the bladder at least 55 gray over four weeks, and mature cartilage up to 70 gray. The lung, a particularly complex organ, is much more sensitive, while subtle, but possibly important, changes can take place in blood vessels at quite low doses.

Of course, therapeutic doses – like any dose – may still give rise to tumours later in life or cause hereditary effects. Therapeutic doses, however, are normally given to treat cancer, where life expectancy is short and patients are usually too old to have much chance of having children. The risk is plainly acceptable. However, the risk of incurring these long-term effects at the very much lower doses normally encountered in life and at work causes much more difficulty to scientists and much more controversy among the general public.

Cancer

Cancer is much the most important effect of low level radiation – at least as far as those actually exposed are concerned. In fact, extensive studies of about 100,000 people who were irradiated by, but survived, the A-bomb explosions at Hiroshima and Nagasaki in 1945, have so far shown cancer to be the only cause of increased mortality.

UNSCEAR relies heavily on the studies of these A-bomb survivors in its attempts to estimate cancer risks. It also uses other research, on the rates of cancer among Pacific islanders contaminated by fall-out from a bomb test in 1954, among uranium miners, and among people who have received radiation therapy. But the Hiroshima and

tolerate very much higher doses.

Unborn children are also particularly vulnerable to brain damage, if their mothers are irradiated between the eighth and fifteenth week of pregnancy. This is the period during which the cortex of the brain is formed, and there is a big risk that radiation, such as from x rays, will cause severe mental retardation. Some 30 children irradiated in the womb when the Hiroshima and Nagasaki bombs exploded were affected in this way. Though the individual risks are high – and the effects of the damage are particularly

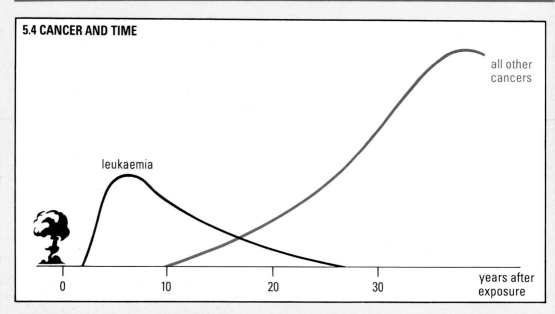

5.4 CANCER AND TIME

Nominal risk of cancer from a single dose of one rad (one hundredth of a gray) uniform whole body irradiation. The diagram, based on research among A-bomb survivors, demonstrates the approximate time of the appearance of malignancies after whole body irradiation. Thus leukaemias appear first, after a latent period of two years, peaking at six to seven years, and tending to disappear after 25 years. Solid tumours begin to appear after 10 years, but there is not yet sufficient information to complete the curve. Diagram from a paper by W. K. Sinclair in the proceedings of the Twentieth Annual Meeting of the National Council on Radiation Protection and Measurements, April 4–5, 1984.

Nagasaki studies are the only ones which, for more than 30 years, have closely followed large numbers of people of all ages who were exposed fairly uniformly to radiation over their whole bodies.

Despite these studies, information on radiation-induced human cancer remains limited. There is a wealth of experimental information from animal tests, but, while it helps, it cannot substitute for evidence of what actually happens to people. For valid estimates to be made of the risks humans face, the human evidence needs to fulfil a whole set of conditions. It must be clear what dose of radiation was absorbed. The dose should have been distributed sufficiently uniformly through the body or, at least, through the particular part of the body being studied. The irradiated population must be watched for decades to give time for all the cancers to emerge. Diagnosis must be good enough to catch all the cancers. It is

particularly important also to have a good "control" population, comparable in every relevant way to the people being studied, except for the fact that they have not been irradiated, to show how many cancers would have occurred in the absence of the radiation. And both populations must be big enough to produce adequate statistics. None of the studies fulfils all these conditions adequately.

More fundamentally still, almost all the evidence results from studying people whose tissues have received quite high doses of radiation, one gray or more. There are only few data about the effects of the range of doses received at work, and no direct information at all about exposures normally received by the general public. So there is no alternative to trying to extrapolate estimates of risks at low doses from what little is known about hazards at high ones.

UNSCEAR – like other bodies working in the field – makes two basic assumptions,

5.5 DOSES AND EFFECTS

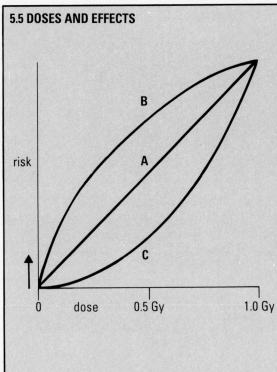

risk

0   dose   0.5 Gy   1.0 Gy

Influence of dose-response relationships on increments of risk for increments of dose. We know roughly the risk of developing cancer from radiation after dose equivalents of one gray as a result of studies of A-bomb survivors and other irradiated populations. We also know, of course, that the risk from zero radiation exposure, were such a thing possible, would be zero. But we know little of the effects of intermediate doses, and so have to try to extrapolate estimates of risks at low doses from what we know about the risks at high ones.

The diagram shows various ways in which this can be done. Very generally speaking, three sorts of graphs could be drawn between the fixed points of zero and one gray (assuming, as UNSCEAR and other bodies do, that there is no threshold, and that therefore any increase in dose will produce an increased risk of cancer, however small). One graph (A) is a straight line, assuming that the risk increases steadily in direct proportion to the dose. The second (B) is convex, suggesting that the risk rises steeply at low doses and more slowly at high ones. The third (C) is concave, suggesting, in contrast, that the risk rises only gradually at low doses and faster at high ones. UNSCEAR, like other bodies working in the field, assumes that the risk increases linearly, that is as suggested by graph A.

which are broadly supported by what evidence there is. The first is that there is no threshold at which there is no risk of cancer. Any dose, however small, increases the chances that the person who receives it will get cancer – and every additional dose will increase those chances further. The second assumption is that the risk increases in direct proportion to the dose; that doubling the dose will double the risk, trebling it will treble it, etc. (diagram 5.5). UNSCEAR believes that this is probably a conservative assumption; that it may overestimate doses at low levels, and will almost certainly not underestimate them. On this, admittedly imperfect, but convenient, basis it is possible to arrive at rough estimates of the risks of different types of cancer.

Leukaemia seems to be the first cancer to emerge in a population after irradiation (diagram 5.4). It seems, on average, to kill about ten years after the original damage was

done – much earlier than other cancers. The numbers of deaths from leukaemia among Hiroshima and Nagasaki survivors dropped sharply after 1970, and it looks as if the toll is now almost complete. So the risk of death from leukaemia can be assessed more confidently than hazards from most other cancers. UNSCEAR estimates that two people out of every thousand will die of leukaemia for every gray they receive. In other words, if someone gets a dose of one gray to red bone marrow there is one chance in 500 that he or she will die of the disease as a result.

Breast and thyroid cancers seem to be the most common tumours caused by radiation. UNSCEAR estimates that around ten people in every thousand will contract thyroid cancer – and ten women in every thousand contract breast cancer – per gray. But both cancers can commonly be cured, and radiation-induced thyroid cancers have particularly low mortality rates. So only about five women in every thousand are likely to die of breast cancer per gray, and only one person in every thousand is expected to perish of thyroid cancer.

Lung cancer, by contrast, is a sure killer. It, too, is a common cancer in irradiated people. Information about rates of lung cancer among uranium miners in Canada, Czechoslovakia, and the United States has emerged to supplement the data from Hiroshima and Nagasaki. Curiously, however, it conflicts with it; even accounting for the different types of radiation involved, uranium miners appear to be between four and seven times more likely to get cancer for every unit of dose than A-bomb survivors. UNSCEAR has considered several reasons for the discrepancy, not least the fact that the miners are generally older than the Japanese population at the time of irradiation. It suggests that five people out of every thousand aged 35 or over at exposure may die from lung cancer for every gray, but only half that number in a population representing all ages. The higher figure is at the lower end of the range indicated by the miners' studies.

Other cancers seem to be less readily induced by radiation. UNSCEAR reckons that only about one person in every thousand is likely to die of cancers of the stomach, liver or large intestine per gray, and that cancers of the bone, oesophagus, small intestine, urinary bladder, pancreas, rectum and lymphatic tissues pose even lower risks, perhaps of about 0.2 to 0.5 per thousand per gray.

Children are more vulnerable than adults, and babies in the womb may be more vulnerable still. Some studies have indicated that children are more likely to die of cancer if their mothers have been x-rayed during pregnancy; but UNSCEAR is not yet convinced that cause and effect have been proved. Japanese babies who were irradiated in the womb at Hiroshima and Nagasaki proved no more likely to get cancer.

There are, in fact, several more discrepancies between the Japanese and other data. Besides the conflict of evidence

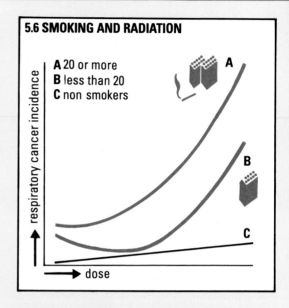

5.6 SMOKING AND RADIATION

A 20 or more
B less than 20
C non smokers

respiratory cancer incidence

dose

Mortality from respiratory cancer related to radon daughter exposure among uranium miners smoking more than 20 cigarettes a day, less than 20 a day, and non smokers.

over lung cancer above, there are substantial discrepancies for both breast and thyroid cancers. In both of these cases, too, the Japanese figures showed a much lower incidence of cancer than the studies elsewhere; in both cases UNSCEAR took the higher figures for its estimates. Such discrepancies illustrate the difficulty of estimating risks at low doses from scanty evidence at high ones. Estimates are made even more difficult by uncertainty over what doses were actually received by the A-bomb survivors. New evidence has called the old calculations of dose into question, and the whole subject is at present being reviewed.

Not surprisingly, when estimates are so difficult, there is considerable debate over how much of a risk there is from low-level radiation. More research is needed. Studies of people exposed to levels of radiation usually encountered in the workplace and the environment would be particularly helpful. Unfortunately the lower the exposure, the harder it is to do meaningful research. It is estimated, for example, that, unless

UNSCEAR's estimates are substantially out, a study of all cancers in nuclear workers exposed to over 0.01 gray a year would need to cover several million person-years to have any hope of coming up with a significant result. And studies of people exposed to environmental levels would be very much harder still.

There are some even more complex issues requiring research. In principle, for example, radiation may interact with other chemical and biological agents to increase cancer rates further. Clearly this is a particularly important issue because radiation is so ubiquitous and because there are so many factors in modern life that could interact with it. UNSCEAR has carried out a preliminary analysis of information on a large number of such factors. Several suspects have emerged, but there is strong evidence on only one – tobacco smoke. Uranium miners seem to get cancer earlier if they smoke (diagram 5.6). In every other area the data are scanty, and more observations are needed.

There have long been suggestions that exposure to radiation may accelerate the aging process and so shorten life. UNSCEAR has also recently reviewed the data on this hypothesis, and has been unable to find enough evidence to substantiate it either in man or in animals, at least at moderate to low long-term exposures. Irradiated populations do have a shorter average life-span, but this seems to be entirely accounted for by the increased number of individuals contracting cancer.

Genetic effects

The study of genetic effects caused by radiation is even more difficult than the study of cancer. This is partly because there is extremely little information on what genetic damage humans sustain through irradiation, partly because the full tally of hereditary conditions takes many generations to show, and partly because, like cancer, these defects would be indistinguishable from those occurring from other causes.

About ten per cent of all babies born alive suffer from some kind of hereditary defect (diagram 5.7, overleaf). These range from mild afflictions like colour blindness to severely incapacitating conditions like Down's syndrome, Huntington's chorea, or severe malformations. Many of the most severely affected embryos and foetuses do not survive; it has been estimated that about a half of all spontaneous abortions have an abnormal genetic constitution. Even if they do survive to birth, babies with hereditary defects are about five times more likely to die before their first birthday than normal children.

Genetic effects fall into two main categories; chromosome aberrations involving changes in the number or structure of chromosomes, and mutations of the genes themselves. The gene mutations split further into dominant mutations (which show in the children of the people who first sustained them), and recessive mutations (which only show up when two people with the same mutated gene jointly conceive a child, and so may lie dormant for many generations, or for ever). Both classes of effects can cause hereditary disease in subsequent generations, but will not necessarily do so. UNSCEAR's estimates concentrate only on severe hereditary conditions.

Only two probable mutations have been found among more than 27,000 children of parents exposed to relatively high doses from the Hiroshima and Nagasaki explosions – and none at all among roughly the same number of offspring of people who received lower exposures. Studies also show no significant increase in the incidence of chromosome abnormalities in children whose parents were irradiated by the bombs. And although some surveys suggest that irradiated parents are more likely to have Down's syndrome babies, other studies do not.

There is some intriguing evidence that people exposed to low doses do suffer detectable chromosome damage in their blood cells. This has been shown at remarkably low levels of exposure in people living at Badgastein, Austria, or working in its supposedly therapeutic radioactive springs. Nuclear workers exposed to less than

5.7 HEREDITARY DEFECTS

current incidence in general population
per million live births

90,000

the internationally accepted maximum permissible level of radiation in the Federal Republic of Germany, the United Kingdom and the United States also exhibit chromosome damage. But the biological significance and health consequences of such damage have not been established.

In the absence of further data, it is necessary to estimate the risks of hereditary defects in man on the basis of extensive tests on animals. UNSCEAR employs two methods of trying to assess the risk to man. One concentrates directly on determining how much damage is done by a given dose of radiation. The other tries to derive what doses are needed to double the number of offspring that would normally be born with hereditary defects of different types.

The first method estimates that one gray of low level radiation administered to males alone will cause between 1000 and 2000 severe mutations, and 30 to 1000 severe effects due to chromosome aberrations, in every million births. The figures for the irradiation of women are much more uncertain, but lower, as female germ cells are less sensitive to radiation; rough calculations suggest that they range from zero to 900 per million births for mutations and from zero to 300 for chromosome aberrations.

The second method estimates that continuous irradiation of one gray per generation (about 30 years) will produce about 2000 severe cases of genetic disease per million births in the children of those exposed. It also attempts to work out the total number of defects that will arise over **all** generations if the same rate of exposure continues. It reckons that ultimately about 15,000 children out of every million will continue to be born with severe disease as a result (diagram 5.7).

This method attempts to include the effects of recessive mutations. Not much is known about them, and they are still a subject of debate; but it is thought that they will only make a minor contribution as the chances of mating between two people with exactly the same kind of gene damage are small. Little is known, either, about the effects of radiation on characteristics such as height and fertility which are determined not by a single gene, but by many acting together. UNSCEAR's estimates concentrate mainly on the effects on single genes because the assessment of the contributions of such polygenic factors is extremely hard.

A greater limitation is the fact that both methods of estimation are only able to address serious hereditary effects. The evidence strongly suggests that minor defects grossly outnumber serious ones, so much so that they could well cause more harm to the general population.

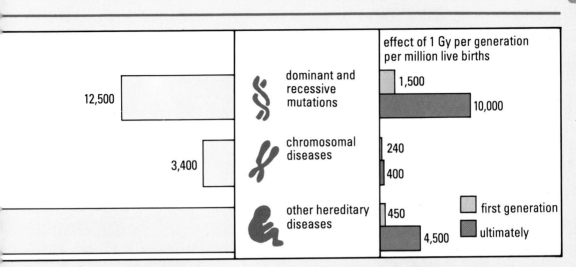

		effect of 1 Gy per generation per million live births
12,500	dominant and recessive mutations	1,500 / 10,000
3,400	chromosomal diseases	240 / 400
	other hereditary diseases	450 / 4,500

first generation

ultimately

Hereditary defects are fairly widespread, but additional doses of radiation can increase their incidence. The diagram shows the current incidence of serious defects in the general population (which will include the effect of natural background radiation) and what UNSCEAR believes would be the additional number of serious defects due to one gray of radiation a generation both in the children of those first exposed, and over all generations if the same rate of exposure continued. The figures are given as the number of children born with the serious defects per million live births.

In its latest report, UNSCEAR made a first attempt to estimate the human impact of serious genetic defects. It tried to assess, and differentiate between, the harm from different genetic defects. For example, both Down's syndrome and Huntington's chorea are serious genetic diseases – but they have a very different impact. Huntington's chorea first strikes between the third and fifth decades of life and causes very serious, but gradual, degeneration of the central nervous system; Down's syndrome causes extremely serious problems from birth onwards. If a distinction is to be made between the two, Down's syndrome could be seen to have the greater impact.

UNSCEAR has, therefore, attempted to evaluate the effects of genetic disease in terms of years of impaired and lost life. This, of course, does not make adequate allowance for the suffering of victims, and can make no allowance at all for such factors as the anguish of the parents of an affected child – but these are impossible to quantify anyway. Conscious that it was only making a first, crude attempt,

therefore, UNSCEAR estimated in its last report that one gray of constant irradiation every generation would give rise to 50,000 years of impaired life, and another 50,000 years of lost life, for every million live births in the children of the first generation irradiated – and ultimately a total of 340,000 years of impaired life and 286,000 years of lost life per every million live births.

Crude though they may be, these estimates are important because they represent an attempt to try to take social values into account when assessing the effects of radiation. For it is such values, in addition to numerical estimates, that are increasingly influencing attitudes towards the acceptability of risks. And it is right that this should be so.

6

This chapter, unlike the preceding four, is not based on the reports of UNSCEAR, but addresses a subject that they have never covered.

At this point, there are good grounds for legitimate confusion. If the assessments reported in this booklet are anything like correct, low-level radiation poses a relatively minor public hazard.

Many people readily accept much greater hazards from, for example, smoking and driving. A citizen of a developed country receiving an average dose from both natural and man-made sources of radiation is five times more likely to die on the road, and more than a hundred times more likely to perish from smoking 20 cigarettes a day, than he or she is to contract a fatal radiation-induced cancer.

There seems to be little public concern about natural radiation, which contributes about four-fifths of the average annual effective dose equivalents world-wide. Few people, for example, seem to move from areas with high background radiation to places offering lower exposures so as to minimize their risk of getting cancer. There is also virtually no public concern about the two human activities that deliver the greatest unnecessary exposures – energy conservation, and overhigh exposures from medical x rays. Almost all public attention and apprehension is focused on nuclear power, one of the smallest contributors to the overall dose.

Scientists and administrators in many countries are often annoyed by what they see as public irrationality – and sometimes even suggest that it is aroused by agitators who want to undermine society itself. This is unwise, as the British Royal Society has pointed out. The public attitude is not as irrational as it seems and may be well founded. And, quite rightly, most Governments will follow public opinion, rather than the recommendations of "experts".

One reason for the gap in perception between the majority of experts and a growing proportion of the public may stem from the very imprecision of assessments of

6.1 RISKS – PERCEIVED AND REAL

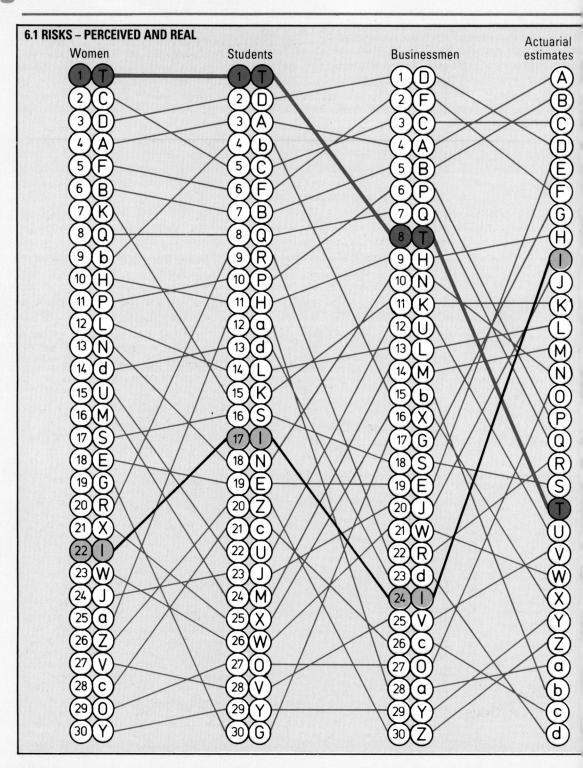

Women | Students | Businessmen | Actuarial estimates

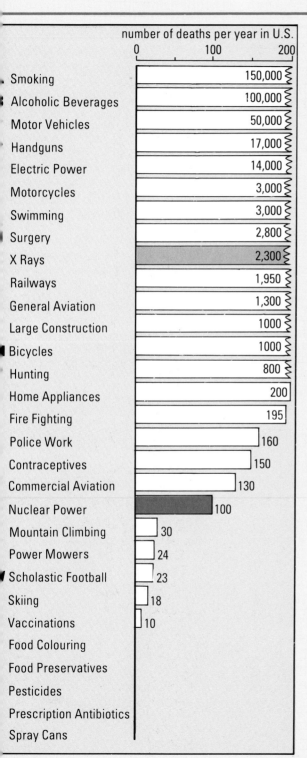

number of deaths per year in U.S.

Smoking	150,000
Alcoholic Beverages	100,000
Motor Vehicles	50,000
Handguns	17,000
Electric Power	14,000
Motorcycles	3,000
Swimming	3,000
Surgery	2,800
X Rays	2,300
Railways	1,950
General Aviation	1,300
Large Construction	1000
Bicycles	1000
Hunting	800
Home Appliances	200
Fire Fighting	195
Police Work	160
Contraceptives	150
Commercial Aviation	130
Nuclear Power	100
Mountain Climbing	30
Power Mowers	24
Scholastic Football	23
Skiing	18
Vaccinations	10
Food Colouring	
Food Preservatives	
Pesticides	
Prescription Antibiotics	
Spray Cans	

the effects of some exposures. This booklet has repeatedly stressed the problems involved in collecting reliable information on some types of radiation exposure and in assessing their effects. Determining the acceptability of risks is far harder still. Little is known about why people react to risks as they do. And the available methods for measuring the costs and benefits of hazardous undertakings are still very imprecise.

As illustrated in the previous chapter, measurements of the cost of disability and disease are still crude. Normally, they only try to quantify the effects of excess mortality, often in financial terms; at best, they attempt some rough assessment of life impairment from gross injury. They cannot effectively value the impact of lesser damage to the quality of life, let alone take account of human distress and frustrated prospects. But the public does take such factors into account, however instinctively.

It is often even harder to assess benefits than to determine costs. Furthermore, it is not sufficient to show that a hazardous process benefits society as a whole; the people most at risk want to be sure that the benefit to **them** outweighs the hazard. In radiation therapy for cancer, the chance of a cure usually far outweighs the risks of the high doses, and the people receiving the doses are those who stand to benefit from them. Unjustified exposures from medical x-ray examinations provide an equally straightforward equation: the patient is being exposed to extra risk for no extra benefit.

The risks that the public perceives to be the highest are not always those that actually kill the most people. Three groups of Americans, members of the League of Women Voters, college students, and members of business and professional clubs were asked to rank 30 hazards. The orders they chose, in the first three columns, are compared with actuarial estimates of the annual contribution of the hazards to the numbers of deaths in the United States. Nuclear power, ranked first by both the women and the students and eighth by the businessmen, comes twentieth on the list of actual risks. X rays, ranked low by all three groups, come ninth on the actuarial table.

Environmental exposure to radiation from nuclear power, however, presents a much more difficult problem to resolve. In the first place, it is society as a whole that enjoys whatever benefits the energy provides: the people living near nuclear facilities, who shoulder almost all the risk, get only a small proportion of the benefit. Secondly, there is genuine debate over whether nuclear energy does provide a net benefit over the use of other fuels – though the main two alternative practices being considered also present risks. Coal burning gives rise to radioactive fly ash and other serious pollutants, and energy conservation presents its own radiation hazards.

Then there is a substantial difference between voluntary and involuntary risks. Some people gladly embrace especially high risks for fun; they find that danger adds to the enjoyment of hang-gliding or ski-jumping, for example. Others cheerfully defy great hazards for altruistic purposes; people regularly risk their lives to save animals they do not even own. Both smoking and driving involve taking voluntary risks, which is one reason why many people find them acceptable.

While the freedom to risk one's life and health is a necessary part of liberty, the freedom to impose such risk on others is not – and public opinion is acutely aware of this. It consistently takes a harsher view of imposed or involuntary risks. When people feel impotent in the face of such a risk, and have no control over it or means of protecting themselves from it, they are even less tolerant. Radiation from the nuclear fuel cycle is seen by the public as embodying all these undesirable characteristics.

Nuclear power also falls foul of a basic moral dilemma. People doubt whether it is right to bequeath radioactive wastes, which will remain dangerous far into the future, to subsequent generations – particularly as their descendants will have no control over the problem left to them, and as the decision of what to do with the wastes is to be made by the same generation that reaps the benefits from nuclear power. It also suffers from its

association with the revulsion people feel for nuclear war.

Furthermore, people fear catastrophes, however infrequent, more than small dangers, however common. Quite correctly, much of the fear of nuclear power is of the consequences of an accident – whether at a reactor, reprocessing plant, or waste disposal facility – rather than of the effects of routine releases of radiation. UNSCEAR does not address the probabilities of accidents, and those studies that have done so have failed to provide wide public assurance.

Attitudes to risk are also affected by the extent to which they are known. On the one hand, some risks are scarcely known to the public at all, and thus, unfortunately, receive little attention: this probably accounts for the lack of concern about radon in houses in most countries and about unnecessary exposure to x rays. On the other hand, familiarity seems to breed contempt. One recent study showed that well-known risks, such as from motorcycle riding, skiing, climbing, smoking – and even muggers and heroin – inspired little fear. Nuclear power, on the other hand, was both one of the least familiar and one of the most feared sources of risk; interestingly it was far more feared than asbestos, which was judged to be better known.

Secrecy – particularly the half-kept secret – fuels fear, and there has been too much of it in the past. There have also been too many bland reassurances and admonitions that the experts know best. The reassurances have been found wanting, and the experts – though undoubtably highly knowledgeable in their own fields – often do not take a wide enough view. There has been a serious loss of credibility.

The public needs to be involved far more in assessing the risks it is being asked to undertake – and in passing judgement on them. Unless it is, it will be increasingly unwilling to accept them. For this purpose it needs full, factual and unemotional information. However, as Alexander Pope put it: "a little learning is a dangerous thing". This booklet has been an attempt to increase it.